Lecture Notes in Artificial Intelligence   3419

Edited by J. G. Carbonell and J. Siekmann

Subseries of Lecture Notes in Computer Science

T0223565

Boi Faltings   Adrian Petcu
François Fages   Francesca Rossi (Eds.)

Recent Advances
in Constraints

Joint ERCIM/CoLogNet International Workshop
on Constraint Solving and Constraint Logic Programming, CSCLP 2004
Lausanne, Switzerland, June 23-25, 2004
Revised Selected and Invited Papers

 Springer

Series Editors

Jaime G. Carbonell, Carnegie Mellon University, Pittsburgh, PA, USA
Jörg Siekmann, University of Saarland, Saarbrücken, Germany

Volume Editors

Boi Faltings
Adrian Petcu
Ecole Polytechnique Federale de Lausanne (EPFL)
School of Computer and Communication Sciences
Institute of Core Computing Science
Artificial Intelligence Laboratory
IN (Ecublens), 1015 Lausanne, Switzerland
E-mail: {boi.faltings,adrian.petcu}@epfl.ch

François Fages
Institut National de Recherche en Informatique et en Automatique - INRIA
Domaine de Voluceau, Rocquencourt, BP 105, 78153 Le Chesnay Cedex, France
E-mail: Francois.Fages@inria.fr

Francesca Rossi
University of Padova
Department of Pure and Applied Mathematics
Via G.B. Belzoni 7, 35131 Padova, Italy
E-mail: frossi@math.unipd.it

Library of Congress Control Number: 2005921905

CR Subject Classification (1998): I.2.3, F.3.1, F.4.1, D.3.3, F.2.2, G.1.6, I.2.8

ISSN 0302-9743
ISBN 3-540-25176-6 Springer Berlin Heidelberg New York

Springer is a part of Springer Science+Business Media

springeronline.com

© Springer-Verlag Berlin Heidelberg 2005
Printed in Germany

Typesetting: Camera-ready by author, data conversion by Olgun Computergrafik
Printed on acid-free paper   SPIN: 11402763   06/3142   5 4 3 2 1 0

Preface

Constraint programming is a very successful fifth-generation software technology with a wide range of applications. It has attracted a large community of researchers that is particularly strong in Europe.

In particular, constraint programming is the focus of the Working Group on Constraints of the European Research Consortium for Informatics and Mathematics (ERCIM) as well as a major interest of the European Network on Computational Logic (CoLogNET). These groups jointly sponsored a workshop on Constraint Satisfaction and Constraint Logic Programming (CSCLP 2004) held June 23–25 at the Ecole Polytechnique Fédérale de Lausanne (EPFL) in Switzerland. It was hosted by the Artificial Intelligence Laboratory of the EPFL, which is also a member of both groups.

This book presents a collection of papers that are either revised and extended versions of papers accepted at the workshop, or were submitted in response to the open call for papers that followed. The 15 papers in this volume were selected from 30 submissions by rigorous peer review.

The editors would like to take the opportunity to thank all authors and reviewers for the hard work they contributed to producing this volume. We also thank ERCIM and CoLogNET for their support of the workshop and the field of constraint programming in general. We hope the reader will find this volume helpful for advancing their understanding of issues in constraint programming.

December 2004

Boi Faltings
Adrian Petcu
François Fages
Francesca Rossi

Organization

This workshop was jointly organized as the 9th Meeting of the ERCIM Working Group on Constraints, coordinated by François Fages, and the 2nd Annual Workshop of the CoLogNET area on Constraint Logic Programming, coordinated by Francesca Rossi.

Organizing Institutes

The organization was handled by the EPFL, INRIA and the University of Padua.

Organizing and Program Committee

Boi Faltings   EPFL, Switzerland
Adrian Petcu   EPFL, Switzerland
François Fages   INRIA, France
Francesca Rossi   University of Padua, Italy

Sponsoring Organizations

Referees

Ola Angelsmark
Alexis Anglada
Roman Bartak
Nicolas Beldiceanu
Claudio Bettini
Stefano Bistarelli
Lucas Bordeaux
Berthe Y. Choueiry
François Fages
Jie Fang
Stephan Frank
Renker Gerrit
Joel Gompert
Venkata Praveen Guddeti
Brahim Hnich
Petra Hofstedt
Alan Holland
Peter Jonsson

Irit Katriel
T.K. Satish Kumar
Arnaud Lallouet
Steve Prestwich
Paraskevi Raftopoulou
Arathi Ramani
Stefan Ratschan
Igor Razgon
Dirk Reckmann
Francesca Rossi
Olga Tveretina
Petr Vilim
Richard Wallace
Toby Walsh
Armin Wolf
Neil Yorke-Smith
Yaling Zheng
Peter Zoeteweij

Table of Contents

GCC-Like Restrictions on the *Same* Constraint

Nicolas Beldiceanu[1], Irit Katriel[2], and Sven Thiel[2]

[1] LINA FRE CNRS 2729, École des Mines de Nantes, FR-44307 Nantes Cedex 3, France
`Nicolas.Beldiceanu@emn.fr`
[2] Max-Planck-Institut für Informatik, Stuhlsatzenhausweg 85, 66123 Saarbrücken, Germany
`{irit,sthiel}@mpi-sb.mpg.de`

Abstract. The *Same* constraint takes two sets of variables X and Z such that $|X| = |Z|$ and assigns values to them such that the multiset of values assigned to the variables in X is equal to the multiset of values assigned to the variables in Z. In this paper we extend the *Same* constraint in a GCC-like manner by adding cardinality requirements on the values. That is, for each value we have a lower and upper bound on the number of variables that can be assigned this value. We show an algorithm that achieves arc-consistency for this constraint and a faster algorithm that achieves bound-consistency for a restricted case of it.

1 Introduction

The $Same(X = \{x_1, \ldots, x_n\}, Z = \{z_1, \ldots, z_n\})$ constraint [2] is defined on two sets X and Z of distinct variables such that $|X| = |Z|$ and each $a \in X \cup Z$ has a finite domain $D(a)$. A solution is an assignment of values to the variables such that the value assigned to each variable belongs to its domain and the multiset of values assigned to the variables of X is identical to the multiset of values assigned to the variables of Z.

This constraint can be used to model simple scheduling problems such as the one described in [2]: The organization Doctors Without Borders [11] has a list of doctors and a list of nurses, each of whom volunteered to go on one rescue mission in the next year. Each volunteer specifies a list of possible dates and each mission should include one doctor and one nurse. The task is to produce a list of pairs such that each pair includes a doctor and a nurse who are available on the same date and each volunteer appears in exactly one pair.

In the setting described above, the number of potential rescue missions on each day is infinite, so we do not care how the doctor-nurse pairs are distributed between the dates. This paper deals with a variant of *Same* which we call *Same With Cardinalities* (*SWC*) and which allows us to model the doctor-nurse problem when for each date there is a minimum number of missions that must be staffed and a maximum number of missions that are possible. The reader should be reminded of the Global Cardinality Constraint (*GCC*) [5, 7, 8, 10], which is defined on one set of variables and specifies for each value the minimum and maximum number of variables that are to be assigned this value.

Formally, the $SWC(X = \{x_1, \ldots, x_n\}, Z = \{z_1, \ldots, z_n\}, C = \{c_{v_1}, \ldots, c_{v_{n'}}\})$ constraint is specified on two sets X and Z, each containing n assignment variables, and a third set C of n' count variables. With each assignment variable $a \in X \cup Z$ we associate a domain

B. Faltings et al. (Eds.): CSCLP 2004, LNAI 3419, pp. 1–11, 2005.

$D(a) \subseteq D = \{v_1, \ldots, v_{n'}\}$. The count variable c_{v_i} refers to $v_i \in D$ and its domain is an interval $D(c_{v_i}) = [L_i, U_i]$. A solution to the *SWC* constraint is an assignment of values to the variables in $X \cup Z$ such that:

- Each $a \in X \cup Z$ is assigned a value in its domain $D(a)$.
- Each c_{v_i} is assigned a value in the interval $D(c_{v_i})$.
- The multiset of values assigned to the variables of X is equal to the multiset of values assigned to the variables of Z.
- The number of variables in X (and hence also in Z) which are assigned the value v_i is equal to the number assigned to c_{v_i}.

In other words, for a tuple $t \in D^n$ and a value $v \in D$, let $occ(v,t)$ be the number of occurrences of the value v in t. Then the set S of all solutions to the constraint is:

$$S = \{(u_1, \ldots, u_n \; ; \; w_1, \ldots, w_n \; ; \; o_1, \ldots, o_{n'}) \; | $$
$$\forall_j \; u_j \in D(x_j) \; \wedge \; \forall_j \; w_j \in D(z_j) \; \wedge$$
$$\forall_i \; occ(v_i, (u_1, \ldots, u_n)) = occ(v_i, (w_1, \ldots, w_n)) = o_i \in D(c_{v_i})\}.$$

1.1 Arc-Consistency and Bound-Consistency

Given a constraint with domains for the variables, the first question is whether $S \neq \emptyset$, which means that there is at least one assignment of values to the variables that satisfies the constraint. The second question is whether there are efficient *filtering* algorithms for this constraint. That is, algorithms that shrink the domains of the variables by removing values that cannot participate in any solution. The *arc-consistency* problem is to reduce the domain of each variable a such that $D(a)$ is the projection of S onto the component that corresponds to a. That is, a value v remains in $D(a)$ iff there is a solution to the constraint in which a is assigned the value v. In the *bound-consistency* problem we assume that the values are linearly arranged, i.e., $v_1 < \cdots < v_{n'}$ and for each $a \in X \cup Z$, $D(a)$ is a contiguous interval of values, i.e., $D(a) = [\underline{D}(a), \overline{D}(a)]$. The problem is to shrink the intervals to the minimum sizes such that $S \subseteq D(x_1) \times \cdots \times D(x_n) \times D(z_1) \times \cdots \times D(z_n) \times D(c_{v_1}) \times \cdots \times D(c_{v_{n'}})$. I.e., the domain of the kth variable is bound-consistent iff S contains at least one tuple whose kth component equals the smallest (largest) value in it.

1.2 $SWC = 2 \times GCC$?

The *SWC* constraint can be modeled by two *Global Cardinality* constraints [5, 7, 8, 10], one on the set X and the other on the set Z, where count variables which are associated with the same value are not duplicated. We show here that consistency for all of the variables of the *GCC* constraints (including assignment and count variables) does not imply consistency for the *SWC* constraint.

In our example, $|X| = |Z| = 2$ and $|Y| = 4$. The domains of the assignment variables are: $D(x_1) = \{1,2\}$, $D(x_2) = \{3,4\}$, $D(z_1) = \{1,2,3,4\}$ and $D(z_2) = \{3,4\}$ and the domain of each count variable is $\{0,1\}$. By examining the variable-value graphs[1] shown in Figure 1, one can easily see that all values are consistent with respect to the two *GCC*

---

[1] This construction will be formally defined in Section 2.

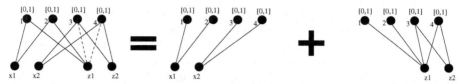

Fig. 1. Example showing that consistency of the two *GCC*'s does not imply consistency of the *SWC* constraint (even when all cardinalities are in [0,1]).

constraints $GCC(\{x_1,x_2\},\{c_{v_1},c_{v_2},c_{v_3},c_{v_4}\})$ and $GCC(\{z_1,z_2\},\{c_{v_1},c_{v_2},c_{v_3},c_{v_4}\})$, but that an arc-consistency or bound-consistency computation for the *SWC* constraint would remove 3 and 4 from the domain of z_1: If z_1 is assigned 3 or 4, then 1 and 2 cannot both be assigned to the same number of variables from X and Z because one of them must be assigned to x_1 and neither can be assigned to z_2.

1.3 Filtering with Flows/Matchings

Network flows were used to design filtering algorithms for several globals constraints. These algorithms follow a general scheme: the constraint is modeled as a network such that there is a correspondence between feasible integral flows in the network and solutions to the constraint. The algorithm finds a feasible flow in this network, constructs the residual graph with respect to this flow and computes the strongly connected components (SCCs) of the residual graph. Then, it is shown how to use the flow and the SCCs to reduce the domains of the variables to arc-consistency or bound-consistency.

Régin was the first to use this approach when he designed an arc-consistency algorithm for the *AllDifferent* constraint [9], which he later generalized for the Global Cardinality Constraint [10]. Mehlhorn and Thiel [6] showed that this scheme gives rise to a faster bound-consistency algorithm for *AllDifferent*. They noticed that in the bound-consistency problem, the network on which the flow and SCC computations need to be performed has a certain structure, *convexity*, which can be exploited in order to perform these computations more efficiently. Katriel and Thiel [5] showed how to exploit convexity to achieve a fast bound-consistency algorithm for *GCC*. Later, the authors [2] defined the *Same* and *UsedBy* constraints and designed arc-consistency and bound-consistency algorithms for them, which also follow the flow-based paradigm. The networks that model the *Same* and *UsedBy* constraints are more complex than the ones used for *AllDifferent* and *GCC*, a fact that also complicates the filtering algorithms, in particular the efficient bound-consistency algorithms. In this paper we show how to model the *SWC* constraint. The network we use resembles the one that was used for the *Same* constraint, but the capacity requirements for the values add a new twist: until now, all networks consisted of a bipartite graph with a node for each value on one side and a node for each variable on the other side, plus two special nodes. The network we use for *SWC* breaks away from this line: each value is modeled by two nodes that are connected by an edge. This structure complicates things even further, in particular in the bound-consistency computation.

1.4 Filtering for the *SWC* Constraint

In the next section we present filtering algorithms for the assignment variables of the *SWC* constraint. The first algorithm achieves arc-consistency and runs in time $O(n^2n')$.

The second algorithm achieves bound-consistency in the restricted case of *SWC* in which $D(c_{v_i}) = [0,1]$ for all $1 \leq i \leq n'$. It runs in time $O(nn')$. As we have noted above, *SWC* is a *GCC*-like restriction of the *Same* constraint. Similarly, the case in which $D(c_{v_i}) = [0,1]$ for all i is analogous to the *AllDifferent* constraint [6, 9, 12] which is the special case of *GCC* in which the capacities for all variables are $[0,1]$.

2 An Arc-Consistency Algorithm for *SWC*

We represent the *SWC* constraint as a flow problem in a directed graph $\vec{G} = (V, \vec{E})$ which we call the *variable-value graph*. The nodes of $\vec{G}$ are $V = \{X \cup Z \cup Y_{in} \cup Y_{out} \cup \{s, t\}\}$ where $Y_{in} = \{y_1^{in}, \ldots y_{n'}^{in}\}$ and $Y_{out} = \{y_1^{out}, \ldots y_{n'}^{out}\}$. In other words, there is a node a for each variable $a \in X \cup Z$, there are two nodes y_i^{in}, y_i^{out} for each value v_i where $1 \leq i \leq n'$ and there are two additional nodes s and t. The edges of $\vec{G}$ are:

- For each $x_j \in X$ and $i \in D(x_j)$, $(x_j, y_i^{in}) \in \vec{E}$ with capacities $[0,1]$.
- For each $z_j \in Z$ and $i \in D(z_j)$, $(y_i^{out}, z_j) \in \vec{E}$ with capacities $[0,1]$.
- For each $1 \leq i \leq n'$, $(y_i^{in}, y_i^{out}) \in \vec{E}$ with capacities $[L_i, U_i]$.
- For each $1 \leq j \leq n$, $(s, x_j) \in \vec{E}$ with capacities $[1,1]$.
- For each $1 \leq j \leq n$, $(z_j, t) \in \vec{E}$ with capacities $[1,1]$.
- $(t, s) \in \vec{E}$ with capacities $[n, n]$.

Table 1. Domains of the variables for our example.

j	$D(x_j)$	$D(z_j)$
1	[1,2]	[2,3]
2	[3,4]	[4,5]
3	[4,6]	[4,5]

Table 2. Domains of the count variables for our example.

i	1	2	3	4	5	6
$[L_i, U_i]$	[0,1]	[1,2]	[0,3]	[1,4]	[0,2]	[0,1]

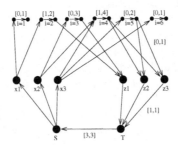

Fig. 2. The variable-value graph for the example in Tables 1 and 2.

Figure 2 shows the graph $\vec{G}$ for the following input. $|X| = |Z| = 3$ and $|Y| = 6$. The domains of the variables of $X \cup Z$ are as in Table 1 and the domains of the count variables are as in Table 2.

The following definition comes from flow theory. See Figure 3 for an example of a feasible flow.

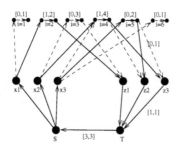

Fig. 3. An integral feasible flow in the graph of Figure 2. The solid edges carry flow and the dashed edges do not.

Definition 1. *Given a directed graph $\vec{G} = (V, \vec{E})$ with lower and upper capacities l_e, u_e for each edge $e \in \vec{E}$, a feasible flow in $\vec{G}$ is a function $f : E \to \mathbb{R}$ such that*

1. **Flow Conservation:** *For each node $v \in V$,*

$$\sum_{\{u|(v,u)\in\vec{E}\}} f(v,u) = \sum_{\{w|(w,v)\in\vec{E}\}} f(w,v).$$

2. **Capacities:** *For each $e \in \vec{E}$, $l_e \leq f(e) \leq u_e$.*

An integral *feasible flow is a feasible flow such that for all $e \in \vec{E}$, $f(e)$ is an integer.*

Lemma 1. *There is a one-to-one correspondence between the integral feasible flows in $\vec{G}$ and the solutions to the constraint.*

Proof. Let f be an integral feasible flow in $\vec{G}$. For each $x \in X$, the amount of flow coming into x (from s) is exactly 1, hence there is exactly one $y \in Y_{in}$ which is connected to x by an edge that carries non-zero flow. Similarly, the flow out of each $z \in Z$ (to t) is exactly 1 so there is exactly one $y \in Y_{out}$ which is connected to z by an edge that carries non-zero flow. For each $a \in X \cup Z$, let $I(a)$ be the index of this node y. That is, $I(a) = i$ such that $y \in \{y_i^{in}, y_i^{out}\}$.

Then we can construct the solution

$$SWC(\{I(x_1), \ldots, I(x_n)\}, \{I(z_1), \ldots, I(z_n)\}, \{f(y_1^{in}, y_i^{out}), \ldots, f(y_{n'}^{in}, y_{n'}^{out})\}).$$

For all $a \in X \cup Z$, $I(a)$ is well defined. Since the edges $(x, y_{I(x)}^{in})$ and $(y_{I(z)}^{out}, z)$ carry flow, they exist in $\vec{G}$, which implies that $I(a) \in D(a)$ for all $a \in X \cup Z$. In addition, by flow conservation and by the choice of capacities for the edges (y_i^{in}, y_i^{out}) we have $L_i \leq |\{x \in X | I(x) = i\}| = |\{z \in Z | I(z) = i\}| \leq U_i$ for all $1 \leq i \leq n'$, so each value is assigned the same number of times to variables of X and Z, and this number is within its capacity requirements. Hence, the constraint is satisfied.

On the other hand, any solution $SWC(\{I(x_1), \ldots, I(x_n)\}, \{I(z_1), \ldots, I(z_n)\}, \{o_1, \ldots, o_{n'}\})$ where $I(a)$ is the value assigned to the variable a, allows us to construct an integral feasible flow f as follows.

- For each $x \in X$, $f(x, y_{I(x)}^{in}) = 1$ and $f(x, y_j^{in}) = 0$ for all $j \in D(x) \setminus I(x)$.
- For each $z \in Z$, $f(y_{I(z)}^{out}, z) = 1$ and $f(y_j^{out}, z) = 0$ for all $j \in D(z) \setminus I(z)$.

- For each $1 \leq i \leq n'$, $f(y_i^{in}, y_i^{out}) = o_i$.
- For each $x \in X$, $f(s,x) = 1$.
- For each $z \in Z$, $f(z,t) = 1$.
- $f(t,s) = n$.

Since $I(a) \in D(a)$ for all a, all the edges through which we wish to pass positive flow exist in the graph. In addition, since $I(a)$ is determined for all variables, we have that $f(t,s)$ is n and $f(s,x) = f(z,t) = 1$ for $x \in X$ or $z \in Z$. Since $L_i \leq |\{x \in X | I(x) = i\}| = o_i = |\{z \in Z | I(z) = i\}| \leq U_i$ for all $1 \leq i \leq n'$, we get that the total amount of flow into y_i^{in} is equal to the total amount of flow out of y_i^{out} and to the amount of flow through (y_i^{in}, y_i^{out}), and that it is within the capacity range of the edge (y_i^{in}, y_i^{out}). Hence the flow is an integral feasible flow. $\qquad\square$

After finding an integral feasible flow f in $\vec{G}$, we construct the residual graph $\vec{G}_f = (V, \vec{E}_f)$. The edges in $\vec{E}_f$ are as follows. An edge between $a \in X \cup Z$ and $y \in Y_{in} \cup Y_{out}$ appears in $\vec{E}_f$ in its original orientation iff it carries flow zero and in its reverse direction iff it carries flow 1. The edge (y_i^{in}, y_i^{out}) exists iff $f(y_i^{in}, y_i^{out}) < U_i$ and the edge (y_i^{out}, y_i^{in}) exists iff $f(y_i^{in}, y_i^{out}) > L_i$. There are no edges touching s and t. Figure 4 shows the residual graph for our example, with respect to the flow of Figure 3. The following lemma states that we can use the residual graph to determine which edges of the graph are consistent.

Lemma 2. *Let $e = (u,v)$ be any edge in $\vec{G}_f$ with $u \in X \cup Z$ or $v \in X \cup Z$. Then $f'(e) = f(e)$ for all feasible flows f' in $\vec{G}$ iff u and v do not belong to the same strongly connected component (SCC) of $\vec{G}_f$.*

Proof. Standard flow theory. $\qquad\square$

Lemma 3. *An edge $e = (u,v) \in \vec{G}_f$ with $u \in X \cup Z$ or $v \in X \cup Z$ is consistent iff $f(e) = 1$ or u and v belong to the same SCC.*

Proof. If $f(e) = 1$ then e participates in the solution that corresponds to the flow f and is therefore consistent. Otherwise, by Lemma 2 we get that there is a flow f' such that $f'(e) = 1$ (and hence a solution that uses the assignment represented by e) iff u and v belong to the same SCC. $\qquad\square$

Lemma 3 implies the last step of the filtering algorithm: For each variable $a \in X$, remove a value i from $D(a)$ if $f(a, y_i^{in}) = 0$ and a, y_i^{in} do not belong to the same SCC

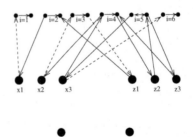

Fig. 4. The residual graph with respect to the flow shown in Figure 3. The dashed edges are not consistent.

of the residual graph. Similarly, for each variable $a \in Z$, remove a value i from $D(a)$ if $f(y_i^{out}, a) = 0$ and a, y_i^{out} do not belong to the same SCC of the residual graph.

Recall that $n = |X| = |Z|$ and $n' = |Y_{in}| = |Y_{out}|$. Clearly, $|\vec{E}| = O(nn')$. The running time of the algorithm is dominated by the time required to find a flow, which is $O(n|\vec{E}|) = O(n^2 n')$ [4, 10]. Hence:

Theorem 1. *There is an algorithm that reduces the domains of the assignment variables of the SWC constraint to arc-consistency. This algorithm runs in time $O(n^2 n')$ where n is the number of assignment variables and n' is the cardinality of the union of their domains.*

3   A Bound-Consistency Algorithm for *SWC* with $[0, 1]$ Cardinalities

In this section we show that bound-consistency can be computed in time $O(nn')$ for the special case of *SWC* in which $[L_i, U_i] = [0, 1]$ for all $1 \leq i \leq n'$. In the bound-consistency setting, the domain of each variable is a contiguous interval of values. This implies that we can construct a variable-value graph which is *convex* [3], i.e., a bipartite graph such that the nodes on one side (the value nodes) can be arranged such that the neighborhood of each variable node is a contiguous sequence of value nodes. As was done with regards to other constraints [2, 5, 6], we will show that the filtering algorithm described above can be implemented faster when the graph is convex. In particular, we will show that an integral feasible flow can be found in time $O(n' + n \log n)$. Thus, the running time of the algorithm is not dominated by the time required to find a flow but rather by the time required to find SCCs in the residual graph, which is linear in the number of edges, i.e., $O(nn')$.

3.1   From Flows to Matchings

As was the case with other constraints, it is convenient in the bound-consistency context to think of matchings in an undirected bipartite graph instead of flows in a directed network. For the *SWC* constraint this is perhaps even more significant than for other constraints, because as will be shown in the following, the fact that we speak of matchings instead of flows allows us to simplify the variable-value graph: Each value i can be represented by just one node y_i instead of the pair of nodes y_i^{in}, y_i^{out}. In particular, this means that the graph is bipartite with a node for each variable on one side and a node for each value on the other side, unlike $\vec{G}$.

The bound-consistency algorithm will operate on the undirected variable-value graph $G = (V, E)$ where $V = X \cup Y \cup Z$ with $Y = \{y_1, \ldots, y_{n'}\}$ and $E = \{(a, y_i) | a \in X \cup Z \wedge i \in D(a)\}$. Before describing the algorithm, we need more definitions.

Definition 2. *Let $M \subseteq E$ be a set of edges of G. For any node $v \in X \cup Y \cup Z$, $N_M(v)$ is the set of nodes which are neighbors of v in $G' = (X \cup Y \cup Z, M)$.*

The following definition appears in [2]:

Definition 3. *Let $M \subseteq E$ be a set of edges of G. We say that M is a* parity matching *in G iff $\forall_{v \in X \cup Z} |N_M(v)| = 1$ and $\forall_{y \in Y} |N_M(y) \cap X| = |N_M(y) \cap Z|$.*

That is, a parity matching is a set of edges that match each node in $X \cup Z$ with exactly one node from Y and each node from Y with the same number of nodes from X and from Z. We add the following definition.

Definition 4. *Let $M \subseteq E$ be a set of edges of G. We say that M is a twins matching in G iff M is a parity matching and $\forall_{y \in Y} |N_M(y) \cap X| = |N_M(y) \cap Z| \in \{0, 1\}$.*

In other words, a twins matching is a parity matching such that for each $y \in Y$, y is either matched with one node from X and one node from Z or it is not matched at all. The following lemma shows the connection between twins matchings in G and integral feasible flows in $\vec{G}$.

Lemma 4. *Let G and $\vec{G}$ be the graphs corresponding to the SWC constraint with $[0, 1]$ cardinalities. Then there is a one-to-one correspondence between the twins matchings in G and the integral feasible flows in $\vec{G}$.*

Proof. Given a twins matching M in G, we can construct an integral feasible flow f in $\vec{G}$ as follows. For an edge $(x_j, y_i) \in E$ with $x_j \in X$ and $y_i \in Y$, $f(x_j, y_i^{in}) = 1$ if $(x_j, y_i) \in M$ and $f(x_j, y_i^{in}) = 0$ otherwise. Similarly, for an edge $(z_j, y_i) \in E$ with $z_j \in Z$ and $y_i \in Y$, $f(y_i^{out}, z_j) = 1$ if $(z_j, y_i) \in M$ and $f(y_i^{out}, z_j) = 0$ otherwise. For an edge $(y_i^{in}, y_i^{out}) \in \vec{E}$, $f(y_i^{in}, y_i^{out}) = |\{x_j | x_j \in X \wedge (x_j, y_i) \in M\}|$. Finally, $f(t, s) = n$ and for all $1 \leq j \leq n$, $f(s, x_j) = f(z_j, t) = 1$.

It is clear that the capacity requirements are met for any edge which is adjacent to at least one node from $\{s, t\} \cup X \cup Z$. For an edge (y_i^{in}, y_i^{out}), the amount of flow is equal to the number of nodes from X that were matched with y_i in M, that is, either 0 or 1, which is also within the capacity requirements of the edge (y_i^{in}, y_i^{out}). It remains to show that flow conservation holds. The flow into and out of each of s and t is of value n, and for a node in $X \cup Z$ it is of value 1. For each $1 \leq i \leq n'$, either no flow enters y_i^{in} and no flow leaves y_i^{out}, in which case y_i was not matched with any node from X and hence also from Z, so $|\{x_j | (x_j, y_i) \in M\}| = 0$ and this implies $f(y_i^{in}, y_i^{out}) = 0$, or a flow of value 1 goes into y_i^{in} and a flow of the same value leaves y_i^{out} because y_i was matched with a node from X and a node from Z. In this case, $f(y_i^{in}, y_i^{out}) = 1$. In both cases, flow conservation holds at y_i^{in} and y_i^{out}.

For the other direction, assume that we are given an integral feasible flow f in $\vec{G}$. Then we can construct a twins matching M in G as follows. For $(x_j, y_i) \in E$ such that $x_j \in X$ and $y_i \in Y$, $(x_j, y_i) \in M$ iff $f(x_j, y_i^{in}) = 1$. For $(z_j, y_i) \in E$ such that $z_j \in Z$ and $y_i \in Y$, $(z_j, y_i) \in M$ iff $f(y_i^{out}, z_j) = 1$. To see that the matching we obtain is a twins matching, note that for each $a \in X \cup Z$ there is exactly one node in $Y_{in} \cup Y_{out}$ which is connected to a by an edge that carries flow 1. Hence a is matched with exactly one node from Y. In addition, since there is a capacity requirement of $[0, 1]$ on the edge (y_i^{in}, y_i^{out}) for each $1 \leq i \leq n'$, each y_i is matched with at most one node from X and at most one node from Z. From flow conservation we get that it is either matched with one node from X and one node from Z or not matched with any node from $X \cup Z$. $\square$

Figure 5 shows the algorithm for finding a twins matching in G. It makes two passes over the y nodes. The first pass is identical to the algorithm for finding a parity matching in [2]. The proof of its correctness appears in that paper. The second pass uses the parity matching found in the first pass to create a twins matching (if one exists).

In order to prove the correctness of this algorithm, we first show that the parity matching generated by the first pass matches $\{y_1, \ldots, y_i\}$ with the minimal possible

```
(* Assumption: X and Z are sorted according to D. *)
(* Pass 1: Find a parity matching in G. *)
Px ← []  (* priority queue containing x nodes sorted by D̄ *)
Pz ← []  (* priority queue containing z nodes sorted by D̄ *)
for i = 1 to n' do
        forall xh with D(xh) = i do Px.Insert xh
        forall zh with D(zh) = i do Pz.Insert zh
        (* Assume that MinPriority of an empty queue is ∞ *)
        while Px.MinPriority = i or Pz.MinPriority = i do
                if Px.IsEmpty or Pz.IsEmpty then report failure
                x ← Px.ExtractMin
                z ← Pz.ExtractMin
                m[x] ← yi
                m[z] ← yi
        end
endfor
if Px.NotEmpty or Pz.NotEmpty then report failure
(* Pass 2: Use the parity matching to create a twins matching. *)
Px ← []  (* priority queue containing x nodes sorted by D *)
Pz ← []  (* priority queue containing z nodes sorted by D *)
for i = n' downto 1 do
        forall xh with m[xh] = i do Px.Insert xh
        forall zh with m[zh] = i do Pz.Insert zh
        if Px.NotEmpty and Pz.NotEmpty then
                x ← Px.ExtractMax
                z ← Pz.ExtractMax
                Match yi with x and z
        endif
endfor
if Px.NotEmpty or Pz.NotEmpty then report failure
```

Fig. 5. Algorithm to find a twins matching in a convex bipartite graph.

number of nodes, for every i. This justifies the fact that in the second pass, which traverses the nodes of Y from $y_{n'}$ to y_1, a node $a \in X \cup Z$ becomes a candidate for matching starting at $y_{m[a]}$ and not $y_{\overline{D}(a)}$.

Lemma 5. *Let m be the parity matching generated by the first pass of the algorithm and let M be any parity matching in G. Then for all $1 \leq i \leq n'$, $|N_M(y_1, \ldots, y_i)| \geq |N_m(y_1, \ldots, y_i)|$.*

Proof. By induction on i. For $i = 0$ the lemma trivially holds. Assume that it holds for all $1 \leq j < i$ but not for i. That is, $|N_M(y_1, \ldots, y_j)| \geq |N_m(y_1, \ldots, y_j)|$ for all $1 \leq j < i$, but $|N_M(y_1, \ldots, y_i)| < |N_m(y_1, \ldots, y_i)|$.

This means that $|N_m(y_i)| > |N_M(y_i)|$. Let $x_j \in X$ and $z_k \in Z$ be a pair of nodes that are matched with y_i in m but not in M. Then by construction of the algorithm, either $\overline{D}(x_j) = i$ or $\overline{D}(z_k) = i$ (or both). Assume, w.l.o.g., that $\overline{D}(x_j) = i$. This implies that $\overline{D}(x_{j'}) = i$ for any matching mate $x_{j'}$ of y_i. Let ℓ be maximal such that $\ell < i$ and y_ℓ is matched with some $x_{\ell'}$ with $\overline{D}(x_{\ell'}) > i$. Then for any $x_{\ell''}$ that was matched with one of $\{y_{\ell+1}, \ldots, y_i\}$, $\underline{D}(x_{\ell''}) > \ell$, because otherwise by convexity, $x_{\ell''}$ was in P_x when $x_{\ell'}$ was extracted so by construction of the algorithm, $\overline{D}(x_{\ell''}) \geq \overline{D}(x_{\ell'}) > i$, contradicting the maximality of ℓ.

We get that all the nodes that were matched by m with $\{y_{\ell+1}, \ldots, y_i\}$ must be matched with $\{y_{\ell+1}, \ldots, y_i\}$. By the induction hypothesis, we know that the nodes of $\{y_1, \ldots, y_\ell\}$ cannot be matched in M with less nodes than they were in m. The combination of these two facts implies that the lemma holds for i.   □

Lemma 6. *If the algorithm in Figure 5 reports failure then there does not exist a twins matching in G.*

Proof. If the algorithm reports failure during or immediately after the first pass then it was proved in [2] that there does not exist a parity matching in the graph, and in particular there does not exist a twins matching.

It remains to show that if the second pass does not match all nodes of $X \cup Z$ then there does not exist a twins matching in the graph. To do this, we show by induction on i that for all i from n' downto 1, the second pass of the algorithm matches $y_i, \ldots, y_{n'}$ with the maximal possible number of nodes.

For $i = n'$, if the algorithm matches $y_{n'}$ then the claim holds. Assume that the algorithm does not match $y_{n'}$ but there is a twins matching in which $y_{n'}$ is matched with $x_j \in X$ and $z_k \in Z$. Then by Lemma 5, we know that $m[x_j] = n'$ and $m[z_k] = n'$. So x_j was in P_x and z_k was in P_z in iteration n' so the algorithm should have matched $y_{n'}$, a contradiction.

For smaller i, we can assume by the induction hypothesis that the algorithm matched $\{y_{i+1}, \ldots, y_{n'}\}$ with the maximum possible number of nodes. Let c_{i+1} be the number of nodes from X (and hence also from Z) that were matched by the algorithm with $y_{i+1}, \ldots, y_{n'}$. Assume that the claim does not hold for y_i. That is, there is a twins matching M that matches $y_i, \ldots, y_{n'}$ with more nodes than the algorithm. This means that M matches $y_{i+1}, \ldots, y_{n'}$ with c_{i+1} nodes from each of X and Z, M matches y_i and the algorithm does not match y_i. From the fact that the algorithm does not match y_i, we get that at least one of P_x and P_z was empty during iteration i of the algorithm. Assume, w.l.o.g., that P_x was empty. Then there are exactly c_{i+1} nodes $x \in X$ with $m[x] \geq i$, and by Lemma 5 this implies that M cannot match $\{y_i, \ldots, y_{n'}\}$ with more than c_{i+1} nodes from each of X and Z, a contradiction.   □

Lemma 7. *If the algorithm in Figure 5 reports success then it constructs a twins matching.*

Proof. If the algorithm reports success then for each node $a \in X \cup Z$, the first pass assigns a value $m[a]$ between 1 and n'. Hence, in the second pass each such a is inserted into the relevant queue and since the queue is empty at the end, a is also extracted from the queue. Since these node are always extracted in a pair that includes an x node and a z node, and in each iteration at most one such pair is extracted, we get that the matching generated is a twins matching.   □

Thus we have obtained:

Theorem 2. *There is an algorithm that narrows the domains of the assignment variables of the SWC constraint to bound-consistency, if all cardinality requirements for the count variables are $[0, 1]$. This algorithm runs in time $O(nn')$ where n is the number of assignment variables and n' is the cardinality of the union of their domains.*

Conclusion and Open Problems

We have extended the *Same* constraint with GCC-like cardinalities for the values. We have shown an algorithm that achieves arc-consistency for the new constraint and a faster algorithm that achieves bound-consistency for the special case in which all cardinalities are $[0, 1]$.

The bottleneck of the bound-consistency algorithm is the SCC computation. For other global constraints, it was shown that convexity can be exploited to speed up the SCC computation and achieve even faster bound-consistency algorithms. In the *SWC* case, however, we were, so far, only able to exploit convexity in the matching step of the algorithm. It is an open problem whether the SCC computation can be performed faster than $O(|E|)$.

Another question is whether the twins matching algorithm can be generalized to the case of arbitrary cardinalities. That is, is there an algorithm that runs in time $O(n' + n \log n)$ and finds a feasible matching in a variable-value graph with arbitrary cardinalities on the values?

References

1. R. K. Ahuja, T. L. Magnanti, and J. B. Orlin. *Network flows: theory, algorithms and applications*. Prentice Hall, Englewood Cliffs, NJ, 1993.
2. N. Beldiceanu, I. Katriel, and S. Thiel. Filtering algorithms for the Same constraint. In *First International Conference on Integration of AI and OR Techniques in Contraint Programming for Combinatorial Optimization Problems (CPAIOR 2004)*, volume 3011 of *Lecture Notes in Computer Science*, pages 65–79, Nice, France, 2004. Springer.
3. F. Glover. Maximum matchings in a convex bipartite graph. *Naval Res. Logist. Quart.*, 14:313–316, 1967.
4. L. R. Ford Jr. and D. R. Fulkerson. *Flows in Networks*. Princeton University Press, 1962.
5. I. Katriel and S. Thiel. Fast bound consistency for the global cardinality constraint. In *Proceedings of the 9th International Conference on Principles and Practice of Constraint Programming (CP 2003)*, volume 2833 of *LNCS*, pages 437–451, 2003.
6. K. Mehlhorn and S. Thiel. Faster Algorithms for Bound-Consistency of the Sortedness and the Alldifferent Constraint. In *Proceedings of the 6th International Conference on Principles and Practice of Constraint Programming (CP 2000)*, volume 1894 of *LNCS*, pages 306–319, 2000.
7. C.-G. Quimper, A. López-Ortiz, P. van Beek, and A. Golynski. Improved algorithms for the *global cardinality* constraint. In M. Wallace, editor, *Principles and Practice of Constraint Programming (CP'2004)*, volume ? of *LNCS*, pages ?–? Springer-Verlag, 2004.
8. C.-G. Quimper, P. van Beek, A. Lopez-Ortiz, A. Golynski, and S. B. Sadjad. An efficient bounds consistency algorithm for the global cardinality constraint. In *Principles and Practice of Constraint Programming*, pages 600–614, 2003.
9. J.-C. Régin. A filtering algorithm for constraints of difference in CSPs. In *Proceedings of the 12th National Conference on Artificial Intelligence (AAAI-94)*, pages 362–367, 1994.
10. J.-C. Régin. Generalized Arc-Consistency for Global Cardinality Constraint. In *Proceedings of the 13th National Conference on Artificial Intelligence (AAAI-96)*, pages 209–215, 1996.
11. Médecins sans Frontières. http://www.doctorswithoutborders.org/.
12. W.J. van Hoeve. The alldifferent constraint: A survey. In *Submitted manuscript. Available from http://www.cwi.nl/wjvh/papers/alldiff.pdf*, 2001.

A Note on Bilattices
and Open Constraint Programming

Arnaud Lallouet

Université d'Orléans – LIFO,
BP 6759 – F-45067 Orléans – France

Abstract. We propose to use bilattice as a constraint valuation structure in order to represent truth and belief at the same time. A bilattice is a set which owns two lattices orderings. They have been used in Artificial Intelligence in order to model incomplete information. We present a framework for Bilattice-valued Constraint Programming which allows to represent incomplete or conflicting information and to combine constraints with a set of operators. It allows to model a variety of situation such as open constraints and the integration of machine learning into constraint programming, reconciliation of divergent opinions in distributed systems or constraint modules in a software engineering perspective.

1 Introduction

Valuation structures for constraints have become popular because they provide control on specific features which are difficult to represent with classical constraints or more generally with first-order logic. For example, in fuzzy CSPs [12], each tuple is given a preference level in the $\mathbb{R}$-interval $[0, 1]$ which represents how likely the tuple belongs to the constraint. Linguistic constructions such as "tall man" or "powerful computer" can be easily represented in this framework. Further, semiring-based CSP [5] provides a more abstract valuation structure able to represent different notions of truth like the one of fuzzy and probabilistic CSPs.

Obviously, these approaches answer the question of the degree of truth, but for some applications, it is needed to model how much we believe in a given assertion concerning a truth value. This situation arises in distributed reasoning when pieces of knowledge coming from multiple sources have to be combined into a single one. But the way of doing this combination depends on the application. Here are a few scenarii in which this phenomenon occurs:

- *Information Gathering over the Internet*: imagine a situation in which web agents are requested to organize a movie night. Some agents may return different programs for the same movie theater due to obsolete information, some agents may be preferred because they have provided more reliable information in the past or because they better fullfil the preferences of the customer.

B. Faltings et al. (Eds.): CSCLP 2004, LNAI 3419, pp. 12–25, 2005.

- *Multi-expert Reasoning*: different expert may give different conclusion starting from the same premises, and their confidence on the subject may be
  taken into account while merging the information.
- *PDA Synchronization*: nomadic applications create divergent copies of the
  same original database. The reconciliation of theses copies can be made
  according different policies: first or last transaction, preference given to the
  PDA or to the PC.
- *Open Constraints*: classical constraints are defined by a set of allowed tuples,
  the other tuples being implicitly considered as false. This is known as the
  Closed World Assumption of database theory. By allowing another value $\perp$
  for tuples, we can model the absence of information about this tuple, which
  is a different notion of being half way between true and false.

All these problems are gaining considerable attention in the constraint community in order to add constraint-based decision support to web or distributed
applications. But it also raises practical and theoretical issues coming from the
field of reasoning with incomplete or inconsistent information. Philosophical and
epistemological concerns have been at the origin of proposals to give answers to
this problem, mainly by giving a representation to partial and unknown states.
A first proposal in this direction is Kleene's three-valued logic [10] **K3** in which
an additional truth value **I** is intended to represent the absence of information.

Belnap introduced a fourth value [3] in order to model inconsistency and
called his valuation structure **Four** (see figure 1). Bilattices have been introduced
by Ginsberg [9] as a elegant framework to represent incomplete and inconsistent
information in logic. This framework has been later extended by Fitting [7] to
model the semantics of logic programs with negation.

In this paper, we propose to define constraints valued in bilattice structures.
We show that this framework has nice possibilities by providing a way to express
incomplete, missing or conflicting information. Some of theses problems have
already been tackled by existing approaches but we believe that bilattices could
offer a nice unifying framework to many extensions of constraint programming
the same way it did for multi-valued logic [9].

The paper is organized as follow. We first recall some definitions and properties about bilattices in section 2. In section 3 we use this structure as valuation
domain for constraints and we give two notions of solution for a constraint problem. Then we give instances of the framework and ideas of applications area in
section 4.

2   Bilattices

We first start with an example introducing **Four** (displayed on figure 1), which
is the simplest bilattice. Let assume that we want to use constraint-based web
agents to inquire on a computer we want to buy. We may have two sites providing
the same constraint, like for example the list of components in the computer, but
possibly with a different meaning. Before the request, we have no information,
which corresponds to the value $\perp$, which may be understood as intermediate
between true and false, since it will be determined in the future. Suppose that

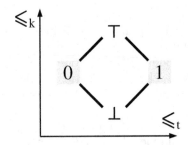

Fig. 1. Four, the bilattice of first-order logic.

we request a 120GB hard drive, it may happen that both site answer "yes, it has one", which can be given the value 1 without ambiguity or both say "no, it does not", which can be given the value 0. A problem arises if one agent says "yes" while the other says "no". If your nature is optimistic, you may combine both informations with a logical 'or', and if you are pessimistic, you may use 'and'. But another sensible point of view is to be skeptical. If we consider truthness, we may think that we are in a situation similar to the original one, intermediate between true and false but for another reason: now we have too much information. This situation is denoted by $\top$.

The new elements $\top$ and $\bot$ may be understood in a set-theoretic way by considering that we collect opinions on a tuple. No opinion is represented by $\emptyset$ which corresponds to $\bot$, false and true correspond to the sets $\{0\}$ and $\{1\}$ respectively, and $\{0, 1\}$ means that we believe that this tuple is both true and false, which is associated to $\top$. Also it is no surprise to consider that classical truth values are in the middle of the knowledge ordering and that absence and excess of information weakens the degree of truth or falsity we may have in an object. Two orderings can be considered for the structure **Four**:

– the *truth* ordering $\leq_t$ in which $0 \leq_t \top \leq_t 1$ and $0 \leq_t \bot \leq_t 1$. In this ordering, $\bot$ and $\top$ are incomparable.
– the *knowledge* or *information* ordering $\leq_k$ in which $\bot \leq_k 0 \leq_k \top$ and $\bot \leq_k 1 \leq_k \top$. In this ordering, 0 and 1 are incomparable.

Both orderings are displayed on figure 1 as a double Hasse diagram in which the vertical direction expresses the degree of knowledge and the horizontal direction the degree of truth. The interesting feature of this structure is that both orderings define complete lattices. Classically, we call $\wedge$ and $\vee$ the operators meet (or glb) and join (or lub) in the truth direction. They correspond to usual conjunction and disjunction. In the knowledge ordering, meet and join are respectively called *consensus* and *gullability* operators [8] and are denoted by $\times$ and $+$.

More generally, a bilattice is composed of a base set equipped with two lattice orderings. If different (otherwise it would collapse to a lattice), these two orderings have to be connected. In the original definition of Ginsberg [9] this connection was made by adding a negation operator for the truth ordering. This operator seems unecessary in the general case for describing the structure itself (although useful for many applications) and therefore has been dropped by Fitting in his definition of *interlaced bilattice* [7]. It follows that a precise definition of the general concept of bilattice is somehow tricky [1]. But for our purpose, we rather use Fitting's notion of *interlaced* bilattice whose properties are better suited for our application to constraints:

Definition 1 (Pre-bilattice).
Let A be a non-empty set containing at least two elements. A pre-bilattice *is a structure $B = (A, \leq_t, \leq_k)$ where $(A, \leq_t)$ and $(A, \leq_k)$ are lattices.*

We denote by $\vee$ and $\wedge$ the lub and glb operations according the truth ordering $\leq_t$ and $+$ and $\times$ the respective ones according to the knowledge ordering $\leq_k$.

Definition 2 (Interlaced bilattice).
An interlaced bilattice *is a pre-bilattice $B = (A, \leq_t, \leq_k)$ such that each of the operations $\vee$, $\wedge$, $+$ and $\times$ is monotonic with respect to both $\leq_t$ and $\leq_k$.*

Bilattices can be complete, in which case there exists a greatest and a smallest element. We denote these elements 1 and 0 for the truth ordering and $\top$ and $\bot$ for the knowledge ordering. They also can be distributive if each operator distributes over the others. Moreover, they have many algebraic properties and we refer to [7] for a more detailed description.

Since bilattice are used in this framework as a valuation structure for constraints, they can be equipped with a notion of negation. Negation is denoted by $\neg$ and reverses the truth ordering and should be monotonic with respect to the knowledge ordering. Hence $\neg\neg a = a$ and $a \leq_k b \Rightarrow \neg a \leq_k \neg b$. This makes sense since we should not be more informed on the negation of a value than on the value itself. Fitting also introduced a similar operation called *conflation*, denoted by $-$ for the knowledge part. Elegant symmetry results follow for bilattices which have both kinds of negation [7].

Lattices can be used to build bilattices as proposed by [9] and [7]. Let $(C, \leq_1)$ and $(D, \leq_2)$ be two (complete) lattices. Then a bilattice $C \odot D = (C \times D, \leq_t, \leq_k)$ can be constructed as follows:

- $(c_1, d_1) \leq_k (c_2, d_2) \Leftrightarrow c_1 \leq_1 c_2$ and $d_1 \leq_2 d_2$.
- $(c_1, d_1) \leq_t (c_2, d_2) \Leftrightarrow c_1 \leq_1 c_2$ and $d_2 \leq_2 d_1$.

The underlying intuition is that, in a couple $(c, d) \in C \odot D$ c measure the belief for the tuple and d the belief against it. Further, [9] has introduced world-based bilattices following the same construction as the instantiation of the underlying lattices by a powerset equipped with set inclusion. Ginsberg [9] and Fitting [7] proved two fundamental representation theorems on bilattices, which were completed by Avron [2]. The first one states that a distributive bilattice is isomorphic to $L \odot L$ where L is a distributive lattice [9, 7], and that an interlaced lattice is isomorphic to $L \odot L$ where L is a bounded lattice [7, 2]. Fitting has further proved that under certain conditions (distributivity of the operators and commutativity of negation and conflation), the consistent members of a bilattice (i.e. $\{e \in A \mid e \leq_k -e\}$) can be represented as intervals of elements of the base set. As an example, consider that $\bot$ in **Four** can be represented by $[0, 1]$ since both values are still possible.

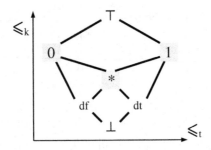

Fig. 2. A bilattice for default reasoning.

It is easy to see that bilattices as valuation domains extend semiring for which only one ordering is required. Moreover, semirings form a lattice structure [5] and can therefore be used to construct bilattices with the above method. However, the extension they provide do not tackle the same knowledge representation problem. Semirings aim at extending the notion of truth we have about an item while bilattices are able to represent truth and confidence at the same time. This makes a fundamental difference for example with what is called EGALITARIANISM AND UTILITARIANISM in [5]. In this approach, a new semiring is built from two others by constructing a specific notion of truth from the two original ones. In particular, there is still an unique ordering in the resulting structure. Moreover, bilattices offer more freedom since they are not always constructed as products of two lattices, like the bilattice for simple default reasoning presented in figure 2, where *df* and *dt* are respectively *default* true and false.

3   Bilattice-Based CSPs

Like semiring-based CSPs, bilattice-based CSPs (or BCSPs) are obtained by replacing the boolean valuation structure of classical CSPs by an arbitrary bilattice. However, since bilattice extend semirings, much of the definitions used in this context are still valid when changing the valuation structure.

Let V be a set of variables and $D = (D_X)_{X \in V}$ their (finite) domains. For $W \subseteq V$, we denote by D^W the set of tuples on W, namely $\Pi_{X \in W} D_X$. Projection of a tuple or a set of tuples on a variable or a set of variables is denoted by $|$. If A is a set, then $\mathcal{P}(A)$ denotes its powerset. For $A \subseteq D^W$, $\overline{A}$ denotes its complement in D^W.

Definition 3 (Constraint).
Let $B = (A, \leq_t, \leq_k)$ *be a bilattice and* $W \subseteq V$. *A constraint is a mapping* $c : D^W \to A$.

In the following, we denote by $var(c)$ the set of variables W of the constraint c. Constraints are composed in order to form CSPs. The most usual composition used in classical CSPs is conjunction but in the context of BCSPs, each operator $\vee, \wedge, +$ and $\times$ of the bilattice can be given a counterpart at the constraint level. Operators at the constraint level are surrounded by a circle.

Definition 4 (Constraint composition).
Let $*$ be an operator on A and c_1 and c_2 two constraints. The composition of c_1 and c_2 according to $*$, denoted by $c_1 \circledast c_2$, is a constraint $c = c_1 \circledast c_2$ such that:

- $var(c) = var(c_1) \cup var(c_2)$
- $c : D^{var(c)} \to A$ *is defined by* $t \mapsto c_1(t|_{var(c_1)}) * c_2(t|_{var(c_2)})$.

All operators are defined for arbitrary composition of constraints of different arities. While it is fairly straightforward for conjunction, the meaning is not obvious for the other ones and we shall consider composition of constraints over the same arities. Let us examine the different operators in turn:

- $\oslash$ is the classical conjunction operator. Most CSPs including the original definition of SCSPs only use this operator.
- $\oslash$ defines a disjunction between constraints. Disjunctive combination of constraints have received little attention so far. In [4], it is used for proving the correctness and completeness of a labeling procedure. More recently, it has been used in [6] to describe gathering of new information for open constraints in a distributed setting. Besides that, there is an objective reason why disjunctive combination had not became popular in the CSP community. This is due to its non-monotonic nature with respect to consistencies. A particular case of disjunctive composition is constraint removal which has been extensively studied under the name of dynamic CSPs [13]. Different methods have been proposed to restore a consistent state after a composition.
- $\oplus$ is a new interesting combinator which allows to register different opinions in a truth-objective manner. In the disjunctive composition, a tuple which was believed as false can become true, but the opposite is not possible. It means that recording one opinion for a tuple will prevail upon 100 opinions against it ! A great reward to optimists ! This composition allows to model a finer shade of reconciliation, to an extent which depends on the underlying bilattice.
- $\otimes$ combines truth values in the direction of maximum consensus, which may yield to consider that having two divergent opinion is the same as not knowing anything.

Of course, no particular composition is better than the others and a suitable use of the operators is application-dependent. Semantics can be given to an arbitrary formula made from these connectives by straightforward induction using definition 4. Here comes the definition of BCSPs:

Definition 5 (BCSP).
A Bilattice-based Constraint Satisfaction Problem (or BCSP) is a tree whose nodes are constraint operators and whose leaves are constraints.

Here follows the definition of solution of such a CSP:

Definition 6 (Solution of a BCSP).
Let C be a BCSP. The solution of C is the constraint defined by the root of the tree.

Classical CSPs only use ⊘. Note that if a BCSP only uses ⊘ operators, then the solution is monotonic with respect to the truth ordering when performing constraint addition. This somehow makes BCSPs an extension of classical CSPs. Of course, anti-monotonicity apply for disjuctive combination in the same context.

As for SCSPs, the notion of solution as a function may be inadequate as it needs to represent large cartesian products and give a value to every tuple. This is why abstract solutions are proposed in [5], which are maximally true solutions. If the lattice is a total order, then the common truth value of all abstract solutions is called the best level of consistency. If not, this best level is an upper bound of the incomparable values of the set of abstract solutions. In BCSPs, we may be interested by maximality according both ordering. For example, assume that we measure independently truth and confidence by a real number in $[0, 1]$, is it more interesting to have a solution with a degree of truth of 1 and a confidence of 0.1 or another one with a truth value of 0.8 but with a confidence of 0.9 ? Actually both are interesting and should be provided to the user.

This can be done by defining a dominance relation between tuples. The most usual dominance relation is due to Pareto and is the basis of multi-objective optimization.

Definition 7 (Pareto dominance).
Let c be a constraint of arity W. The Pareto dominance relation *between tuples of D^W is defined by $t \preceq t' \Leftrightarrow t \leq_t t'$ and $t \leq_k t'$.*

Abstract solutions are non-dominated tuples:

Definition 8 (Abstract solutions).
Let C be a BCSP over the set of variables W. The set AS of abstract solutions of C is the set of non-dominated solutions, i.e. $AS = \{t \in D^W \mid \not\exists t' \in D^W, t \preceq t'\}$.

An implementation of this framework will require the same techniques as for SCSPs, and deserves further investigations. In particular, existing problems to represent approximations and to compute consistencies will be the same. But since the composition of constraints is not only conjunction, it is the notion of consistency itself which causes problems. The classical notion of consistency is that a superset of solutions is maintained and contracted as new information is deduced. But this contraction is monotonic only with respect to further conjuctive composition. Hence the way approximations may compose according to various operators is an interesting open question. The implementation of **Four** is in progress as Open CSPs [11].

4   Instances of the Framework and Applications

In this section, we present some useful bilattices which can be used in presence of incomplete information and some examples of situation in which BCSPs may be useful.

4.1   Open Constraints

The simplest non-trivial bilattice is **Four** depicted in figure 1. While SCSPs are able to model classical CSPs, the simplest bilattice yet provides an extension by allowing to handle incomplete or contradictory information. The new truth values $\perp$ and $\top$ can be used to introduce a new type of constraints with partial definition we call *open constraints*:

Definition 9 (Open constraint, def 1).
An open constraint c is a constraint on the bilattice **Four** $= (\{\perp, 0, 1, \top\}, \vee, \wedge, +, \times, 0, 1, \perp, \top)$.

As noticed earlier, the base set $\{\perp, 0, 1, \top\}$ of **Four** is isomorphic to $\mathcal{P}(\{0, 1\})$ and thus we can give another more intuitive definition of open constraints:

Definition 10 (Open constraint, def 2).
An open constraint c of arity W is a couple (c^+, c^-) of subsets of D^W.

The positive and negative part correspond to the tuples whose truth value is known to be true or false (or both for $c^+ \cap c^-$), and the other ones (which belong to $\overline{c^+ \cup c^-}$) are unknown. Constraints which have no tuple valued to $\top$ are consistent, the others are paraconsistent:

Definition 11 (Paraconsistent constraint).
An open constraint (c^+, c^-) is paraconsistent if $c^+ \cap c^- \neq \emptyset$.

Paraconsistent constraints can deal with local inconsistencies which may appear in distributed reasoning with the following advantages:

- Dealing explicitly with paraconsistency allows to circumsize the potential problems of inconsistency. A computation which does not use paraconsistent tuples may be developped to its end without compromising the correctness of the solution. As an example, having a web site which saying that Elvis is alive and another saying he is dead would probably not bother a deduction on Coltrane's music.
- Even if we want to use a $\top$-valued tuple for a deduction, we can treat it as true or false (according to our level of optimism).

Let $c_1 = (c_1^+, c_1^-)$ and $c_2 = (c_2^+, c_2^-)$. Both orderings can be written in a set-theoretic form:

- truth ordering: $c_1 \leq_t c_2 \Leftrightarrow c_1^+ \subseteq c_2^+$ and $c_1^- \supseteq c_2^-$. In this ordering, $c_1 \leq_t c_2$ if the truth value of every tuple is less in c_1 than in c_2.
- knowledge ordering: $c_1 \leq_k c_2 \Leftrightarrow c_1^+ \subseteq c_2^+$ and $c_1^- \subseteq c_2^-$. In this ordering, $c_1 \leq_k c_2$ if c_1 gives a truth value to less tuples than c_2.

Consistent Open Constraints
Let us first consider only consistent values. Open constraints represent what we know about a relation: the positive part contains tuple which are known as true, the negative part the ones which are known as false. But some other tuples are

simply unknown. Usually in Constraint Programming, a (closed) constraint is defined by giving the set of its allowed tuples[1] (in extension or intentionnaly by some expression in a constraint language whose semantics is well defined). The other tuples are simply completed by the so-called *Closed World Assumption* familiar in database theory: what is not explicitely true is considered as false. Although fine in many contexts, this certainly does not model our actual knowledge which is partial and sometimes contradictory.

An open constraint represents the known part of an hidden real-world relation. It can be viewed as the set of closed constraints which are compatible with its positive and negative parts. But when we want to draw conclusions from our knowledge, we have to choose one of these compatible constraints and bet it actually represent the world. We call such a compatible closed constraint an *extension* of the open constraint. Let us define a classical constraint c by a couple (W, T) where $W = var(c) \subseteq V$ is its *arity* and $T = sol(c) \subseteq D^W$ is its *solution*.

Definition 12 (Extension of a consistent constraint).
Let $c = (c^+, c^-)$ be a consistent open constraint. A classical constraint $c' = (W, T)$ is an extension *of c if $c^+ \subseteq T$ and $c^- \subseteq \overline{T}$.*

In general, many extension can be considered, and let us introduce three of them. Among all possible extensions lies the real constraint which is associated to the real world problem. In most cases, its knowledge is impossible and all can be done is computing an approximation of it. But we recall that computing this approximation is crucial if we want to use this constraint in a classical CSP, and therefore be able to build a solver for it. Let $c = (c^+, c^-)$ be an open constraint. We denote an extension of c by $[c]$:

- *Cautious* extension: $[c]_{\text{cautious}} = (W, \overline{c^-})$. All unknown tuples are assumed to be true (figure 3b). A solver generated according to this extension is cautious in the sense that it will not prune the search space for any unknown tuple.
- *Brave* extension: $[c]_{\text{brave}} = (W, c^+)$. All unknown tuples are assumed to be false (figure 3c). A solver generated according to this extension will prune

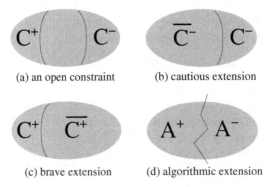

(a) an open constraint   (b) cautious extension

(c) brave extension   (d) algorithmic extension

Fig. 3. An open constraint and some of its extension.

---

[1] Or conversely by giving forbidden tuples.

the search space as soon as possible. Actually, it behaves exactly as for a closed constraint for which all non-allowed tuples are disallowed.

– *Algorithmic* extension $[c]_{\mathcal{A}}$: let $\mathcal{A} : D^W \to \{0,1\}$ be a tuple classification algorithm such that $t \in c^+ \Rightarrow \mathcal{A}(t) = 1$ and $t \in c^- \Rightarrow \mathcal{A}(t) = 0$. Then $[c]_{\mathcal{A}} = (W, \{t \in D^W \mid \mathcal{A}(t) = 1\})$ (figure 3d). This last kind of extension will be the ideal host for a learning algorithm. Tuples from c^+ and c^- are respectively positive and negative examples which are used to feed the learning algorithm. Note that the two preceding extensions are particular cases of this one for constant functions. The main challenge is to be able to generate the best possible solver: the one which has a strong pruning power and is not subject to too many incorrectness errors. As a learning task, there is no universal solution for every problem and the user has to carefully choose and tune his/her learning algorithm in order to obtain good results. It may happen that the algorithm does not classify correctly some tuples defined by c, for noise tolerance for example. In this case, we say that $[c]_{\mathcal{A}}$ defines an *incorrect extension* of c.

In such a way, open constraints can wrap learning algorithms in the same spirit global constraint do for specialized Operation Research algorithms. This framework is developed in [11].

General Open Constraints
If we allow paraconsistent constraints, we have to find a similar way of constructing solvers and integrating them in a CSP. We can give a modified definition of extension which classifies both unknown and paraconsistent tuples:

Definition 13 (Extension).   *[generalization of def 12]*
Let $c = (c^+, c^-)$ be an open constraint. A classical constraint $c' = (W, T)$ is an extension of c if $c^+ \backslash c^- \subseteq T$ and $c^- \backslash c^+ \subseteq \overline{T}$.

Interactions with the external world are modeled by the composition of open constraints. In order to do this, we propose to reformulate the generic composition framework given in definition 4 for our set-theoretic definition of open constraint. Note that these operators have not been designed in a software engineering perspective but rather as building blocks on top of which more evolved operators can be built. They may be user-unfriendly, in particular when dealing with paraconsistency. These operators are defined for constraints on the same set of variables. An extension to any set of variable is easy to imagine.

Operators Associated to $\leq_t$:

– disjunction: $c_1 \otimes c_2 = (c_1^+ \cup c_2^+, c_1^- \cap c_2^-)$.
– conjuction: $c_1 \otimes c_2 = (c_1^+ \cap c_2^+, c_1^- \cup c_2^-)$.
– negation: $\neg c = (c^-, c^+)$ (Ginsberg's negation).

Operators Associated to $\leq_k$:

– gullability: $c_1 \oplus c_2 = (c_1^+ \cup c_2^+, c_1^- \cup c_2^-)$.
– consensus: $c_1 \otimes c_2 = (c_1^+ \cap c_2^+, c_1^- \cap c_2^-)$.
– conflation: $-c = (\overline{c^-}, \overline{c^+})$.

These operators can be used to build CSPs from basic open constraints as in the modularity paragraph below or as revision operators in dynamic CSPs (see [13] for description and pointers). In the latter case, it is not difficult to see that, except $\oslash$, they are all non-monotonic with respect to the classical arc-consistency approximation. Thus dedicated techniques of dynamic CSPs have to be extended for the new operators.

4.2   An Import Operator

In a software engineering perspective, the composition operators provide many ways to combine constraints in order to model a problem. Hence it could be useful to provide an import operator for open constraints (we restrict here to consistent open constraints). We introduce the operator $c_1 \triangleleft c_2$ with the following meaning: tuples are imported by c_1 from c_2 and their truth value in c_2 is prioritary. In other words, the information of c_1 is extended by tuples of a more trusted source c_2 and the value a tuple has in c_1 is kept only if the tuple is unspecified in c_2. Note that $\bot$ is not considered when another value is present since it models the absence of information. The truth value of a tuple in $c_1 \triangleleft c_2$ is summarized in the following table:

c_1	c_2	$c_1 \triangleleft c_2$
0	0	0
0	$\bot$	0
0	1	1
$\bot$	0	0
$\bot$	$\bot$	$\bot$
$\bot$	1	1
1	0	0
1	$\bot$	1
1	1	1

To this end, we propose to define the import operator as follows:

Proposition 14.
Let c_1 and c_2 be two consistent open constraints. Then

$$c_1 \triangleleft c_2 = (c_1 \otimes -c_2) \oplus c_2$$

The proof is done by checking all cases in the truth table of each operator.

Proposition 15.
If c_1 and c_2 are consistent, then so is $c_1 \triangleleft c_2$.

Of course, an operator in which c_1 has precedence over c_2 can be easily defined. Since truth values can be combined by the elementary operators of the lattice, many more operators can be defined. Such operators could bring modularity in constraint programming and reasoning.

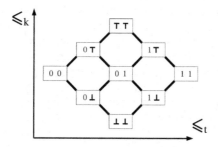

Fig. 4. A bilattice for 2-sources non-ordered reconciliation.

4.3 Reconciliation of Multiple Agents Opinions

When dealing with multiple agents, each one could propose its own opinion about a tuple. This problem is of particular interest in the context of web services which are currently normalized (http://www.w3.org/2002/ws/). Reconciliation of the two points of view requires to define which operator will be used and the underlying bilattice. In this example, we propose to use the $\oplus$ operator which adds information coming from the two sources (but any other could be used according to the application).

In figure 4, the two sources are considered as unordered and any source which contributes with a 1 raises the degree of truth and one which contributes with a 0 lowers this level. The upper part of the bilattice corresponds to the case of one or both sources answer 0 and 1 at the same time.

In figure 5, the same situation is depicted except that the two sources are now ordered. It is supposed that the opinion of the first source is more reliable than the one of the second source. More reliable means here that if the first source says 1, this opinion has greater credit than the same opinion coming from the second source. This can be particularly noticed on states where the sources have divergent opinions.

Example 16.
A company wants to hire a collaborator. The candidate should know either C++ or Java and should have good skills in networks. The project manager wants to

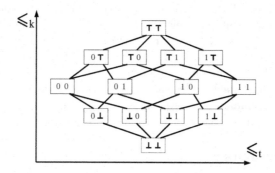

Fig. 5. A bilattice for 2-sources ordered reconciliation.

Table 1. Data retrieved from two job-finding services.

C++		Java		Network	
Mr. Green	$1\perp$	Mr. Green	01	Mr. Green	$0\perp$
Mrs. White	11	Mrs. White	11	Mrs. White	$1\perp$
Mr. Blue	$0\perp$	Mr. Blue	$\perp 0$	Mr. Blue	$\perp 1$
Mr. Red	$\perp 0$	Mr. Red	$\perp 1$	Mr. Red	11
Mrs. Pink	$\perp\perp$	Mrs. Pink	$\perp\perp$	Mrs. Pink	$\perp\perp$

Table 2. Solution of the BCSP.

Name	A = C++ $\otimes$ Java	A $\wedge$ Network
Mr Green	11	$0\perp$
Mrs. White	11	$1\perp$
Mr. Blue	$\perp\perp$	$\perp\perp$
Mr. Red	$\perp 1$	$\perp 1$
Mrs. Pink	$\perp\perp$	$\perp\perp$

use two web-based compagnies: `Job-Online.com` and `FunnyJobs.com`. He thinks that the first company is more serious and therefore uses the 2-sources ordered reconciliation bilattice depicted in figure 5. The web services return data given in table 1 for the three constraints C++, Java and Network. The BCSP can be represented as the following tree:

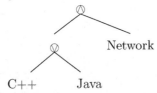

Network

C++   Java

BCSPs, as SCSPs give a truth value to every possible tuple. The results are given in table 2.

5 Conclusion

In this paper, we have proposed bilattices as a new kind of valuation domain for CSP. Besides a classical truth ordering, bilattices propose to add a knowledge ordering able to represent the belief in a particular truth value. We present a framework for Bilattice-valued Constraint Programming which allows to represent incomplete or conflicting information and to combine constraints with a set of operators. It allows to model a variety of situation such as open constraints and the integration of machine learning into constraint programming, reconciliation of divergent opinions in distributed systems or constraint modules in a software engineering perspective.

Acknowledgements

The author is supported by CNRS grant 2JE095.

References

1. Arnon Avron. A note on the structure of bilattices. *Mathematical Structures in Computer Science*, 5(3):431–438, 1995.
2. Arnon Avron. The structure of interlaced bilattices. *Mathematical Structures in Computer Science*, 6(3):287–299, 1996.
3. Nuel D. Belnap. A useful four-valued logic. In J. M. Dunn and G. Epstein, editors, *Modern uses of multiple-valued logic*, pages 8–37. Reidel Publishing, 1975.
4. Stefano Bistarelli, Philippe Codognet, Yan Georget, and Francesca Rossi. Labeling and partial local consistency for soft constraint programming. In Enrico Pontelli and Vítor Santos Costa, editors, *Practical Aspects of Declarative Languages*, volume 1753 of *LNCS*, pages 230–248, Boston, MA, USA, 2000. Springer.
5. Stephano Bistarelli, Ugo Montanari, and Francesca Rossi. Semiring-based constraint satisfaction and optimization. *Journal of the ACM*, 44(2):201–236, March 1997.
6. Boi Faltings and Santiago Macho-Gonzalez. Open constraint satisfaction. In Pascal van Hentenryck, editor, *International Conference on Principles and Practice of Constraint Programming*, volume 2470 of *LNCS*, pages 356–370, Ithaca, NY, USA, Sept. 7 - 13 2002. Springer.
7. Melvin Fitting. Bilattices and the semantics of logic programming. *Journal of Logic Programming*, 11:91–116, 1991.
8. Melvin Fitting. The family of stable models. *Journal of Logic Programming*, 17:197–225, 1993.
9. Matthew L. Ginsberg. Multivalued logics: a uniform approach to reasoning in artificial intelligence. *Computational Intelligence*, 4:265–316, 1988.
10. Stephen C. Kleene. *Introduction to Metamathematics*. Van Nostrand, 1952.
11. Arnaud Lallouet, Andreï Legtchenko, Eric Monfroy, and AbdelAli Ed-Dbali. Solver learning for predicting changes in dynamic constraint satisfaction problems. In Ken Brown Chris Beck and Gérard Verfaillie, editors, *Changes'04, International Workshop on Constraint Solving under Change and Uncertainty*, Toronto, CA, 2004.
12. Thomas Schiex. Possibilistic constraint satisfaction problems or "How to handle soft constraints?". In Didier Dubois and Michael P. Wellman, editors, *Conference on Uncertainty in Artificial Intelligence*, pages 268–275, Stanford University, Stanford, CA, 1992. Morgan Kaufmann.
13. Gérard Verfaillie and Narendra Jussien. Dynamic constraint solving, 2003. CP'2003 Tutorial.

Pruning by Equally Constrained Variables

Igor Razgon and Amnon Meisels*

Department of Computer Science,
Ben-Gurion University of the Negev,
Beer-Sheva, 84-105, Israel
{irazgon,am}@cs.bgu.ac.il

Abstract. We introduce a notion of equally constrained variables of a constraint network. We propose a method of pruning that uses the notion. We combine the proposed method of pruning with FC-CBJ and call the resulting algorithm FC-CBJ-EQ. Our experimental results show that FC-CBJ-EQ outperforms FC-CBJ on constraint networks that encode randomly generated instances of graph k-coloring and of the subgraph isomorphism problems.

1 Introduction

When a problem is modeled as a constraint satisfaction problem (CSP), the obtained model not always expresses specific features of the problem. Recognizing these "implicit" features could significantly reduce the search space. Examples to the fact that recognizing problem-specific features increases efficiency of constraint processing include: symmetry breaking methods (for example, [4, 5, 11]), recognizing tractable classes of CSPs (for example, [1, 3, 14]), methods that detect interchangeability (for example, [6, 15, 2]).

The present paper introduces a method of pruning for complete constraint solvers. The method is based on the discovery of *equally constrained variables*. Informally speaking, two variables v_1 and v_2 of a CSP are equally constrained with a variable v_3 if they are connected to v_3 by the same constraint.

Equally constrained variables occur in CSP-encodings of graph-theoretic problems. Consider, for example, the graph k-coloring problem. Let G be a graph which we would like to color by at most k colors. The problem can be modeled as a CSP in which variables correspond to the vertices of G, all domains are $\{1, \ldots, k\}$, and variables that correspond to adjacent vertices are "connected" by the inequality constraint. Variables v_1 and v_2 are equally constrained with a variable v_3 if the vertices that correspond to v_1 and v_2 are both adjacent to the vertex corresponding to v_3 or both non-adjacent to it.

We propose a modification of FC-CBJ [9] called FC-CBJ-EQ that utilizes information about equally constrained variables. The proposed algorithm differs from FC-CBJ in the following two aspects.

---

* The authors would like to acknowledge the Lynn and William Frankel Center for Computer Sciences for financial support.

B. Faltings et al. (Eds.): CSCLP 2004, LNAI 3419, pp. 26–40, 2005.

1. At the preprocessing stage FC-CBJ-EQ finds all triples of variables v_1, v_2, v_3 of the processed constraint network such that v_1 and v_2 are equally constrained with v_3.
2. During the run of the search algorithm, whenever FC-CBJ-EQ deletes a value val from the current domain of a variable v, it associates with the value a so-called r-set, which is a set of unassigned variables responsible for discarding val. To compute r-sets, we adopt the technique described in [12, 13]. After computation of the r-set of val, FC-CBJ-EQ scans all unassigned variables. If it detects that v and some unassigned variable u are equally constrained with all variables of the r-set associated with val, the value val is removed from the current domain of u.

The proposed method of pruning is closely related to the notion of Neighborhood Partial Interchangeability (NPI) [2] and to the method of Symmetry Breaking via Dominance Detection (SBDD) [4, 5].

As described above, after discarding the value val of a variable v, FC-CBJ-EQ removes val from the domain of every unassigned variable u that is equally constrained with v with respect to every variable of the r-set R of val. In other words, the values val of variables v and u are *interchangeable* with respect to R. Although the formulation of NPI requires the interchangeable values to belong to the domain of the same variable, it is easy to generalize this notion. The proposed method of pruning can be considered as an application of the method of NPI.

Two properties make FC-CBJ-EQ similar to SBDD. First, both algorithms utilize the same pruning approach according to which a node of the search tree can be removed if some "similar" node has already been removed. Second, they check similarity by comparison of the domains of variables. The major difference between FC-CBJ-EQ and SBDD is the detection of r-sets by FC-CBJ-EQ. In contrast, SBDD performs comparison for all the variables. In addition, SBDD uses symmetry, while FC-CBJ-EQ does not.

FC-CBJ-EQ and FC-CBJ are compared on CSPs that encode randomly generated instances of the graph k-coloring and a restricted version of the subgraph isomorphism problems. Our experiments demonstrate that FC-CBJ-EQ outperforms FC-CBJ on these problems.

The rest of the paper is organized as follows. Section 2 provides the necessary background. Section 3 introduces the notion of equally constrained variables and proves the properties that are used to develop the proposed pruning method. Section 4 describes the algorithm FC-CBJ-EQ in detail. Experimental evaluation of the proposed algorithm is presented in Section 5. Finally, Section 6 discusses further development of the proposed approach.

2   Preliminaries

A binary *constraint network* (CN) $Z = \langle V, D, C \rangle$ is a triple consisting of a set of *variables* V, a set of *domains* D and a set of *constraints* C. Let $V = \{v_1, \ldots, v_n\}$. Then $D = \{D_{v_1}, \ldots, D_{v_n}\}$, where D_{v_i} is the domain of values of v_i, $C = \{C_{v_i, v_j} | i \neq j, 1 \leq i, j \leq n\}$, where $c_{v_i, v_j} \subseteq D_{v_i} \times D_{v_j}$ is the set of all compatible pairs of values of v_i and v_j. We refer to the parts of Z as $V(Z)$, $D(Z)$, and $C(Z)$. To emphasize that a value val belongs to the domain of a variable v, we refer to this value as val^v.

Throughout in the paper we illustrate CNs as they are shown on Figure 1 (see, for example CN Z), where the ellipses represent the domains of values of variables and incompatible values of pairs of variables are connected by arcs.

An *assignment* of a CN Z is a pair $\langle v_i, val \rangle$ such that $v_i \in V(Z)$, $val \in D_{v_i}$. A consistent set of assignments is a *partial solution* of Z. A partial solution that assigns all the variables of Z is a *solution* of Z. A CN that has no solution is *insoluble*. A *subnetwork* of a CN Z is a CN obtained by removing variables or values from Z. We define two types of subnetworks.

Definition 1. *A projection of a CN Z to a set of variables $V' \subseteq V(Z)$ denoted by $Z(V')$ is a subnetwork of Z obtained by removing from Z all variables of $V(Z) \setminus V'$.*

Definition 2. *A subnetwork of a CN Z induced by a partial solution P denoted by $Z|_P$ is obtained by removing from Z all the variables assigned in P and removing from the domains of the remaining variables all values inconsistent with P.*

The notions of projection and induced CSP are illustrated on Figure 1.

Definition 3. *A nogood of a CN Z is a partial solution of Z that cannot be extended to a full solution. In other words, P is a nogood of Z if and only if $Z|_P$ is insoluble.*

3   Equally Constrained Variables and Responsibility Sets

Definition 4. *Let Z be a CN and $v_1, v_2, v_3 \in V(Z)$. v_1 and v_2 are equally constrained with v_3 if for every $val' \in D_{v_1} \cap D_{v_2}$ and for every $val'' \in D_{v_3}$, $\langle v_1, val' \rangle$ and $\langle v_3, val'' \rangle$ are compatible if and only if $\langle v_2, val' \rangle$ and $\langle v_3, val'' \rangle$ are compatible[1].*

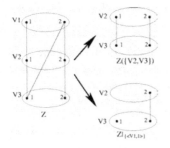

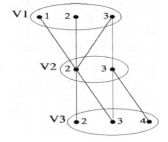

Fig. 1. Illustration of the projection and induced CSP notions.

Fig. 2. Illustration of the notion of equally constrained variables.

---

[1] Note that if $D_{v_1} \cap D_{v_2} = \emptyset$ then v_1 and v_2 are equally constrained with any other variable, though this case is useless for the proposed method of pruning.

Consider the CN on Figure 2. Variables V_1 and V_3 are equally constrained with V_2. This is easy to verify by following the arcs from values 2, 3 of V_3 and V_1 to the values of V_2.

Definition 5. *Let Z be an insoluble CN. A set V' is called a responsibility set of Z if $Z(V' \cap V(Z))$ is insoluble.*

Let Z be a CN and $v \in V(Z)$. Assume that $\{\langle v, val \rangle\}$ is a nogood and let $V' \subseteq V(Z)$ be a responsibility set of $Z|_{\{\langle v,val \rangle\}}$. We formulate for this situation two properties that we use in the next section to construct our pruning procedure.

Proposition 1. *Let $u \in V(Z) \setminus V'$ such that v and u are equally constrained with every variable of V'. If $val \in D_u$ then $\{\langle u, val \rangle\}$ is a nogood and V' is a responsibility set of $Z|_{\{\langle u,val \rangle\}}$.*

To illustrate Proposition 1, consider the CN in Figure 3. We can see that $\{\langle V_1, 1 \rangle\}$ is a nogood and $\{V_2, V_3\}$ is a responsibility set of $Z|_{\{\langle V_1,1 \rangle\}}$. Observe also that V_1 and V_4 are equally constrained with V_2 and V_3. Therefore, $\{\langle V_4, 1 \rangle\}$ is also a nogood with a responsibility set $\{V_2, V_3\}$.

Proof of Proposition 1. Consider two CNs: $Z(V' \cup \{v\})|_{\{\langle v,val \rangle\}}$ and $Z(V' \cup \{u\})|_{\{\langle u,val \rangle\}}$. Since u and v are equally constrained with all variables of V', these two CNs are equal. The former is insoluble, therefore the latter is also insoluble. The insolubility of $Z|_{\{\langle u,val \rangle\}}$ follows from the insolubility of $Z(V' \cup \{u\})|_{\{\langle u,val \rangle\}}$.

Note that V' is a responsibility set of $Z(V' \cup \{u\})|_{\{\langle u,val \rangle\}}$ because V' is the set of variables of the latter, therefore V' is a responsibility set of $Z|_{\{\langle u,val \rangle\}}$. □

Proposition 2. *Let $u \in V'$ such that u and v are equally constrained with every variable of $V' \setminus \{u\}$. Assume that $val \in D_u$ and that for every value $val'(v)$, if $\langle u, val \rangle$ is compatible with $\langle v, val' \rangle$, then $val' \in D_u$ and $\langle v, val \rangle$ is compatible with $\langle u, val' \rangle$. Then $\{\langle u, val \rangle\}$ is a nogood and $V' \setminus \{u\} \cup \{v\}$ is a responsibility set of $Z|_{\{\langle u,val \rangle\}}$.*

Figure 4 illustrates Proposition 2. $\{\langle V_1, 1 \rangle\}$ is a nogood in the CN shown in the figure. A responsibility set of $Z|_{\{\langle V_1,1 \rangle\}}$ is $\{V_2, V_3, V_4\}$. The variables V_1 and V_2 are equally constrained with V_3 and V_4. Also $\langle V_2, 1 \rangle$ is compatible with $\langle V_1, 2 \rangle$ as well as $\langle V_1, 1 \rangle$ is compatible with $\langle V_2, 2 \rangle$. Therefore $\{\langle V_2, 1 \rangle\}$ is a nogood with the responsibility set $\{V_2, V_3, V_4\}$.

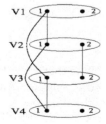

Fig. 3. Illustration of Proposition 1.

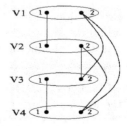

Fig. 4. Illustration of Proposition 2.

Proof of Proposition 2. Assume that the statement of the lemma is not true. Then there is a partial solution P if Z assigning the variables of $V' \cup \{u, v\}$ and such that $\langle u, val \rangle \in P$. Let val' be the assignment of v in P. Let $P' = P \setminus \{\langle u, val \rangle \langle v, val' \rangle\}$.

According to the statement of the lemma, $val' \in D_u$, $\langle v, val \rangle$ is compatible with $\langle u, val' \rangle$ and $\{\langle v, val \rangle \langle u, val' \rangle\}$ is consistent with P'.

Therefore, $P' \cup \{\langle v, val \rangle \langle u, val' \rangle\}$ is a partial solution contradicting the statement that V' is a responsibility set of $Z|_{\{\langle v, val \rangle\}}$. $\qquad\square$

4   A Pruning Procedure for FC-CBJ

In this section we propose a pruning procedure based on the discovery of equally constrained variables. The procedure can be combined with a complete constraint solver. We show its combination with FC-CBJ [9]. We call the obtained solver FC-CBJ-EQ.

The description of the FC-CBJ-EQ solver is divided into three parts. In the first part the maintenance of responsibility sets for insoluble CNs visited by a constraint solver is described. In the second part the algorithm FC-CBJ-EQ is presented. In the last part the correctness of the backtrack procedure of FC-CBJ-EQ is proved.

4.1   Maintaining Responsibility Sets

In this section we describe a method that associates with every value removed from the current domain of a variable a set which we call the r-set of this value. The method is described for FC-CBJ.

FC-CBJ has two reasons to remove values. The first one is that a value of an unassigned variable is incompatible with the last assignment of the current partial solution. Such a value is eliminated by the lookahead procedure of FC-CBJ. Such a value is associated with the empty r-set.

The second reason for removing a value in FC-CBJ is that the domain of some unassigned variable becomes empty. In this case the backtrack procedure discards an assignment of the current partial solution and removes the value of the assignment from the current domain of its variable. The r-set associated with such a value is computed as follows. Let $\langle v, val \rangle$ be the assignment being discarded. Let w be the unassigned variable whose empty current domain causes execution of the backtrack procedure. We associate with val, which is removed from the current domain of v, the set $(S \cup \{w\}) \setminus \{v\}$, where S is the union of the r-sets of all the values of w.

To demonstrate the method consider the following example.

Example 1. Let Z be the CN shown on Figure 5. Assume that FC-CBJ is processing the CN. Let $\{\langle V_1, 1 \rangle, \langle V_2, 1 \rangle\}$ be the current partial solution. After application of the lookahead procedure, the domain of variable V_{100} becomes empty. All the removed values are associated with empty sets. Next, FC-CBJ backtracks. It removes value 1 from the current domain of V_2 and, according to the method described above, associates this eliminated value with the r-set $\{V_{100}\}$.

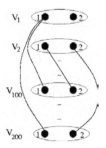

Fig. 5. Illustration of responsibility sets.

The next partial solution tried by FC-CBJ is $\{\langle V_1, 1 \rangle, \langle V_2, 2 \rangle\}$. This partial solution empties the domain of variable V_{200}. Every value of V_{200} is associated with the empty r-set, therefore the r-set associated with $\langle V_2, 2 \rangle$ is $\{V_{200}\}$. When FC-CBJ proceeds, it finds that the current domain of V_2 is empty. It discards the assignment $\langle V_1, 1 \rangle$, removes 1 from the current domain of V_1, and associates with this value the set $\{V_{100}, V_{200}, V_2\}$.

It is possible to see from the example that the r-set of a value is a responsibility set of the CN induced by a nogood that was discarded by removing the value. For example, $\{V_{100}\}$ is a responsibility set of $Z|_{\{\langle V_1, 1 \rangle, \langle V_2, 1 \rangle\}}$, $\{V_2, V_{100}, V_{200}\}$ is a responsibility set of $Z_{\{\langle V_1, 1 \rangle\}}$. We prove a more general claim in Section 4.3.

4.2   Pruning Procedure of Algorithm FC-CBJ-EQ

In this section we describe the algorithm FC-CBJ-EQ. It differs from FC-CBJ in the following three aspects.

1. FC-CBJ-EQ has a preprocessing procedure which computes whether v_1 and v_2 are equally constrained with v_3 for every triple of different variables (v_1, v_2, v_3) of the processed CN. The computation directly follows Definition 4. That is, for every value $a \in D_{v_1} \cap D_{v_2}$ and for every value $b \in D_{v_3}$, FC-CBJ-EQ checks whether $a(v_1)$ and $a(v_2)$ have the same compatibility with $a(v_3)$.
2. The lookahead procedure of FC-CBJ-EQ associates with the empty r-set every value incompatible with the current assignment.
3. FC-CBJ-EQ has a modified backtrack procedure.

The rest of the section describes the backtrack procedure of FC-CBJ-EQ. Similarly to FC-CBJ, FC-CBJ-EQ applies the backtrack procedure when the current domain of some unassigned variable becomes empty.

An important notion used in the algorithm is a *conflict set* [9]. The conflict set of a variable x denoted by $Conf(x)$ is a set of variables assigned by the current partial solution such that if P' is the subset of the current partial solution assigning $Conf(x)$ then for any $val'(x)$ removed from the current domain of x, either P' is incompatible with $\langle x, val' \rangle$ or $P' \cup \{\langle x, val' \rangle\}$ is a nogood.

Algorithm 1 THE BACKTRACK PROCEDURE OF FC-CBJ-EQ.

1: $Conf \leftarrow Conf(w)$
2: Run the backtrack procedure of FC-CBJ (see [9], Section 3.9, the procedure fc-cbj-unlabel)
3: Let the r-set of val^v be equal to $(S \cup \{w\}) \setminus \{v\}$, where S is the union of the r-sets of all the values of w
4: Discard the r-sets of all values that were restored in their current domain executing line 2
5: Let V^* be the r-set of val^v
6: Let V' be the set of all unassigned variables contained in V^*
7: Let A be the subset of assigned variables of $V^* \setminus V'$
8: **for** every unassigned variable u **do**
9:   **if** u and v are equally constrained with all variables of V' and val belongs to the current domain of u **then**
10:     **if** $u \notin V'$ **then**
11:       Remove val from the current domain of u
12:       $Conf(u) \leftarrow Conf(u) \cup Conf \cup A \setminus \{v\}$
13:       Let the r-set of $val(u)$ be equal to V'
14:     **else**
15:       **if** for every val' of the current domain of v compatibility of $\langle u, val \rangle$ with $\langle v, val' \rangle$ implies that val' belongs to the current domain of u and $\langle v, val' \rangle$ is compatible with $\langle u, val' \rangle$ **then**
16:         Remove val from the current domain of u
17:         $Conf(u) \leftarrow Conf(u) \cup Conf(v) \cup A \setminus \{v\}$
18:         Set the r-set of $u(val)$ equal $V' \setminus \{u\} \cup \{v\}$
19:       **end if**
20:     **end if**
21:   **end if**
22: **end for**

Let w be the variable with empty current domain that caused FC-CBJ-EQ to apply the backtrack procedure. Let v be the last (chronologically) assigned variable of $Conf(w)$. Let val be the assignment of v in the current partial solution. Given the data, the backtrack procedure of FC-CBJ-EQ is described in Algorithm 1.

The pseudocode can be divided into three parts. In the first part (lines 1–7), the algorithm keeps the conflict set of w (line 1), deletes val from the current domain of v (line 2), and prepares for further pruning (lines 3–7). In the second part (lines 8–22), the algorithm checks every unassigned variable u and if the current domain of u contains val, the algorithm tries to remove val from the domain. The attempt of removing is described in lines 9–21.

The pseudocode in line 9–21 can again be divided into two parts (lines 9–13 and 15–19, respectively). The first part tries to prune $\langle u, val \rangle$ checking the condition stated in Proposition 1. The second part does the same regarding Proposition 2. If $\langle u, val \rangle$ is pruned in one of these parts, it is associated with a r-set and the conflict set of u is updated. Consistency of updating of these structures is proved in Section 4.3.

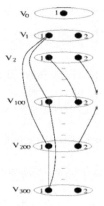

Fig. 6. Illustration of pruning that uses equally constrained variables.

Let us see an example that illustrates a possible scenario of execution of FC-CBJ-EQ when it exhibits more powerful pruning ability than that of FC-CBJ.

Example 2. Let Z be the CN illustrated on Figure 6. Assume that FC-CBJ-EQ is processing Z, its current partial solution is $\{\langle V_0, 1\rangle, \langle V_1, 1\rangle, \langle V_2, 1\rangle\}$, and the current domains of variables V_0, V_1, V_2 equal their initial domains. After a number of backtracks, FC-CBJ-EQ discards all values of V_2 and also $1(V_1)$. As in the previous example, the r-set associated with $1(V_1)$ is $\{V_{100}, V_{200}\}$. Then the backtrack procedure of FC-CBJ-EQ detects that V_1 and V_{300} are equally constrained on both V_{100} and V_{200} and that 1 belongs to the current domain of V_{300}. Therefore V_{300} satisfies the conditions stated in line 9 of Algorithm 1. Further, it satisfies the condition stated in line 10. Therefore, $1(V_{300})$ is deleted from the current domain of V_{300} and associated with the responsibility set $\{V_2, V_{100}, V_{200}\}$.

When FC-CBJ-EQ proceeds to execute it tries to assign V_1 with 2. The lookahead procedure applied after the assignment removes 2 from the current domain of V_{300}. Thus, the current domain of V_{300} becomes empty and FC-CBJ-EQ backtracks. Note that FC-CBJ would not backtrack in this case, because it would not remove $1(V_{300})$ together with removing $1(V_1)$. Next, FC-CBJ tries to reassign V_0, but 1 is the only value in the current domain of V_0, therefore FC-CBJ-EQ reports that the CN is insoluble.

4.3   Correctness

To prove correctness of the backtrack procedure of FC-CBJ-EQ (Algorithm 1), we define a notion of a *consistent state* of FC-CBJ-EQ. Then we show that application of Algorithm 1 to a consistent state of FC-CBJ-EQ produces another consistent state.

Consider FC-CBJ-EQ that is processing a CN Z. A *state* of FC-CBJ-EQ contains the following data:

- the current partial solution;
- the current domains of variables;

- the conflict sets of variables;
- values removed from the current domains of variables and the r-sets associated with them.

Let val^v be a value removed from the current domain of a variable v. Let P' be a subset of the current partial solution assigning the variables of $Conf(v)$, the conflict set of v. We say that $Conf(v)$ is *consistent with* $v(val)$ if either $\langle v, val \rangle$ is incompatible with P' or $P' \cup \{\langle v, val \rangle\}$ is a nogood.

We say that $Conf(v)$ is *consistent* if it is consistent with every value removed from the current domain of v.

The r-set of val^v is consistent if either it is empty or it is a responsibility set of $Z|_{P' \cup \{\langle v, val \rangle\}}$.

A state of FC-CBJ-EQ is *consistent* if all conflict sets and all r-sets are consistent.

Now we are ready to formulate the theorem that claims the correctness of Algorithm 1.

Theorem 1. *Application of Algorithm 1 to a consistent state I of FC-CBJ-EQ produces another consistent state R of FC-CBJ-EQ.*

Proof. We assume that FC-CBJ is correct. Thus, we prove correctness only for those parts of the resulting state R that have been obtained by additional operations performed by FC-CBJ-EQ. In particular, we prove that:

1. the r-set associated with val^v (line 3 of Algorithm 1) is consistent;
2. the conflict sets of all variables whose domains have been pruned in lines 8–22 of Algorithm 1 are consistent;
3. the r-sets of all values removed in lines 8–22 of Algorithm 1 are consistent.

Assume by contradiction that the first claim does not hold. Let V^* be an r-set of val^v computed by Algorithm 1. Inconsistency of V^* means the following.

- V^* is not empty.
- Let P be the subset of the current partial solution that assigns $Conf(v)$ and let $P' = P \cup \{\langle v, val \rangle\}$. Then V^* is not a responsibility set of $Z|_{P'}$.

Note that P' is a subset of the current partial solution at the input state I. Furthermore, P' assigns all the variables of $Conf$, the conflict set of the variable w in state I whose empty current domain causes FC-CBJ-EQ to backtrack. Let P'' be the subset of P' assigning the variables of $Conf(w)$ only.

Considering that the current domain of w is empty in state I, we conclude that P'' is a nogood in Z. Taking into account that $P'' \subseteq P'$ and that V^* is not a responsibility set of $Z|_{P'}$, we get that V^* is not a responsibility set of $Z|_{P''}$. Therefore $Z(Conf \cup V^*)$ has a solution P^* such that $P'' \subseteq P^*$.

Now, recall that V^* consists of w and the union of all r-sets associated with the values of w (Section 3.1) and excludes v. Let $\langle w, val'' \rangle \in P^*$. Let V'' be the r-set of $(val'')^w$ in the state I. Let P^{restr} be the subset of P^* that assigns $Conf \cup \{w\} \cup V'' \setminus \{v\}$. Note that $P'' \subset P^{restr}$.

According to our assumption, the state I is consistent, therefore the r-set V'' of $val''(w)$ is consistent. This means that either it is empty or it is a responsibility set of $Z|_{P'' \cup \{\langle w, val'' \rangle\}}$. In the first case, $\langle w, val'' \rangle$ is incompatible with some assignment of P''. In the second case, $P'' \cup \{\langle w, val'' \rangle\}$ is a nogood in $Z(Conf \cup \{w\} \cup V'')$. In both cases we get a contradiction to the statement that P^{restr} is a partial solution of $Z(Conf \cup \{w\} \cup V'')$. That is, P^* cannot be a partial solution of $Z(Conf \cup V^*)$.

Let us prove consistency of conflict and r-sets produced in lines 9–13. Proving the first part, we got that V^* is not only a responsibility set of $Z|_{P'}$ but also a responsibility set of $Z|_{P''}$. Let A be a subset of V^* that contains variables assigned in the current partial solution but unassigned in P'' (line 7 of Algorithm 1). Let P^A be the subset of the current partial solution of the state R that assigns the variables of A. The set V' obtained in line 6 of Algorithm 1 is a responsibility set of $Z|_{P'' \cup P^A}$.

Considering that $\langle v, val \rangle \in P'' \cup P^A$, we can rewrite $Z|_{P'' \cup P^A}$ as follows: $(Z|_{(P'' \cup P^A) \setminus \{\langle v, val \rangle\}})|_{\{\langle v, val \rangle\}}$. To make the notation shorter, denote the CN in the brackets by Z'.

Observe that $\{\langle v, val \rangle\}$ is a nogood in Z' and that V' is a responsibility set of $Z'|_{\{\langle v, val \rangle\}}$. Therefore if u and v are equally constrained on all variables of V' (line 9) and $u \notin V'$ (line 10) then $\{\langle u, val \rangle\}$ is a nogood in Z' and V' is a responsibility set of $Z'|_{\{\langle u, val \rangle\}}$.

Recall that Z' is a subnetwork of Z induced by $(P'' \cup P^A) \setminus \{\langle v, val \rangle\}$. Therefore $(P'' \cup P^A) \setminus \{\langle v, val \rangle\} \cup \{\langle u, val \rangle\}$ is a nogood in Z. Taking into account that $(P'' \cup P^A) \setminus \{\langle v, val \rangle\}$ is a subset of the current partial solution, we get that $\{\langle u, val \rangle\}$ is inconsistent with the current partial solution. Moreover, when we remove val^u from the current domain of u, we can preserve consistency of $Conf(u)$ by adding the set of variables assigned in $(P'' \cup P^A) \setminus \{\langle v, val \rangle\}$. But the set of variables equals $Conf \cup A \setminus \{v\}$, exactly what is added in line 12. Therefore, the conflict set of u obtained as result of pruning in lines 9–13 is consistent.

The consistency of conflict and r-sets created in lines 15–19 can be proved in a similar way.   □

5   Experimental Evaluation

The goal of our experiments is to evaluate FC-CBJ-EQ on CSPs with a potentially large number of occurrences of equally constrained variables. Such CSPs either contain a small number of different constraints or have some tight constraint that occurs frequently. We selected two types of CSPs that satisfy this property: CSPs that encode the graph k-coloring problem and CSPs that encode a restricted version of the subgraph isomorphism problem where two graphs have the same number of vertices (we refer to the problem as "restricted subgraph isomorphism").

The proposed algorithm could be tested also on randomly generated CSPs, but such testing requires a special generator of random problems that takes into account the properties of the proposed method of pruning. The construction of such a generator is discussed in the last section of the present paper.

The computational effort of CSP search is measured by the number of nodes visited, the number of consistency checks, and the CPU time. Both FC-CBJ and FC-CBJ-EQ were implemented in Microsoft Visual C++ 6.0. The experiments were performed on a computer with Windows 98 Operating System, 300 MHz AMD processor, and 128MB RAM.

The reported measures are the average of 50 runs. In all the experiments both FC-CBJ and FC-CBJ-EQ use the fail first variable ordering heuristic [8] which selects the variable with the smallest domain size and the min-conflict value ordering heuristic [7].

5.1   Graph k-Coloring Problems

Given a graph G and a natural number k. The task is to color G in at most k colors. A CSP Z suitable to this problem is constructed as follows:

- $V(Z)$ corresponds to the set of vertices of G;
- the domain of every variable is $\{1, \ldots, k\}$;
- for any two nonadjacent vertices of G, the corresponding variables are not constrained;
- for any two adjacent vertices, the corresponding variables are connected by the inequality constraint.

FC-CBJ and FC-CBJ-EQ were tested on randomly generated instances of k-coloring with a given number of vertices, colors, and the density of the graphs. The density of a graph is the probability that a pair of vertices of the graph are adjacent.

The results of the experiments are given in Figure 7. The three diagrams from top to bottom present respectively the number of consistency checks performed by FC-CBJ and FC-CBJ-EQ, the number of partial solutions (nodes) visited and the CPU time (in seconds) spent by the algorithms. The parameters of the tested problems are presented along the horizontal axes. Every parameter is represented by a triplet (a, b, c), where a means the number of vertices, b denotes the number of colors, and c is the density. Values of computational effort are represented by columns, black for FC-CBJ and grey for FC-CBJ-EQ. Note that the vertical axes have a logarithmic scale.

The set of parameters for the experiments was selected to cover a wide range of graph densities (10%–80%) and for each density several values of k (the number of colors) were selected. For each pair (colors, density), the size of the graph was selected large enough to take a sizeable computational effort, but still return in a reasonable time.

It is clear that FC-CBJ-EQ outperforms FC-CBJ in the number of nodes visited and the number of consistency checks. With respect to time of run, FC-CBJ-EQ is better for all densities except 30% and 40%.

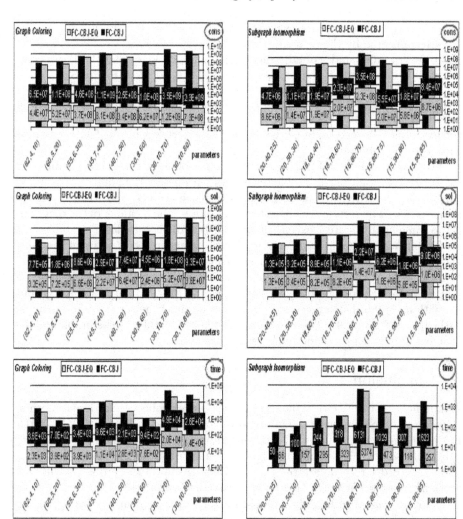

Fig. 7. Evaluation results for the graph k-coloring problem.

Fig. 8. Evaluation results for the subgraph isomorphism problem.

5.2 Restricted Subgraph Isomorphism Problem

The restricted subgraph isomorphism problem consists of two graphs G_1 and G_2 with the same number of vertices. The task is to determine whether G_2 is a subgraph of G_1.

The restricted subgraph isomorphism problem is represented as a CSP Z as follows. The variables of Z are denoted by $V_1, \ldots, V_n$. The values of every domain of Z are denoted by $\{1, \ldots, n\}$. The constraints are defined by the following two rules:

- for every pair of nonadjacent vertices i and j of G_2, the variables V_i and V_j are connected by the inequality constraint;
- for every pair i and j of adjacent vertices of G_2, $\langle V_i, k \rangle$ and $\langle V_j, l \rangle$ are compatible if and only if the vertices $k \neq l$ and k is adjacent to l in G_1.

It is easy to verify that the CSP Z has a solution if and only if G_2 is a subgraph of G_1. Any solution of Z is a mapping from vertices of G_2 to the vertices of G_1. A vertex i of G_2 is mapped to a vertex k of G_1 if V_i is assigned with k in the solution.

FC-CBJ and FC-CBJ-EQ were tested on randomly generated instances of the subgraph isomorphism problem using as parameters the number of vertices, d_1, and d_2. The other two parameters are the densities of G_1 and G_2, respectively (in percents). The results are presented in Figure 8. The meaning of the diagrams is the same as in Figure 7 except the parameters of the problem.

Graphs with different number of vertices were generated to arrive at the same features as that of the graph k-coloring problem.

FC-CBJ-EQ behaves on the subgraph isomorphism problem worse than on graph k-coloring problem. It performs better than FC-CBJ only for dense graphs G_1 with $d_1 \geq 0.6$. For larger densities it can be much better (see the instances with $d_1 = 0.8$ and 0.9). The conclusion is that FC-CBJ-EQ is suitable for the problem only when the underlying graphs are dense.

6   Conclusion

A modification of FC-CBJ called FC-CBJ-EQ that collects additional information during search and uses the information for pruning was presented. FC-CBJ-EQ is shown to outperform FC-CBJ on graph k-coloring and on the restricted subgraph isomorphism problems. These families of problems were selected for testing, in order to evaluate the proposed algorithm on the problems with potentially many occurrences of equally constrained variables.

Another goal of testing the proposed algorithm is to characterize a "threshold" such that "above" it there are problems for which the proposed method of pruning saves computational effort. Such a threshold can be obtained using parameters that determine "uniformity" of the generated CSPs. The expected behavior of FC-CBJ-EQ is that on the most "uniform" CSPs, it would save much computational effort. Then it would be less and less useful as the CSPs being processed become more and more "chaotic". At some moment it must be the case that the additional pruning does not decrease even the number of nodes visited. The parameters of the generator at that moment would constitute the required "threshold". Note that the random CSPs generated given the number of variables, the domain size, density and tightness [10] would probably be "on the wrong side" of the threshold as all the generated constraints would be mutually distinct, therefore no pair of equally constrained variables could be detected.

We pointed out in the introduction that the proposed method of inference closely relates to the notion of NPI and to the method of SBDD. Combination with the approaches is possible further development of the proposed method.

For example, it is not hard to generalize the proposed method of pruning by defining pairs of *values* (not variables) equally constrained regarding sets of variables. Then after the value *val* of v is pruned, it is possible to remove all the values equally constrained with *val* of v on all the variables of R.

It would be interesting to combine SBDD with detection of r-sets. For example, instead of removing values that are interchangeable with *val* of v with respect to a particular set R, one can remove values that are interchangeable with *val* of v with respect to any set that can be mapped to R by some predefined symmetry. Such a strategy being applied to CSPs with many automorphisms could cause huge savings of computational effort.

Another possible way of further development is making FC-CBJ-EQ applicable to nonbinary constraints. We believe that FC-CBJ-EQ can be modified to be applicable to cumulative constraints. The intuition is that any additional nogood pruned by FC-CBJ-EQ differs from an already discovered nogood only in the last assignment. Moreover, the last assignment of the additional nogood has the same value as the last assignment of the known nogood but assigned to another variable. Therefore, every value in the new nogood is assigned to the same number of variables as in the existing nogood. Thus they relate to the cumulative constraint in the same way.

References

1. A. Bulatov and P. Jeavons. An algebraic approach to multi-sorted constraints. In *Principles and Practice of Constraint Programming-CP2003*, pages 183–187, Kinsale, Ireland, oct 2003. Springer.
2. B. Choueiry and G. Noubir. On the computation of local interchangeability in discrete constraint satisfaction problems. In *AAAI/IAAI*, pages 326–333, 1998.
3. M. Cooper, D. Cohen, and P. Jeavons. Characterising tractable constraints. *Artificial Intelligence*, 65:347–361, 1994.
4. T. Fahle, S. Schamberger, and M. Sellmann. Symmetry breaking. In *CP2001*, pages 93–108. Springer, November 2001.
5. F. Focacci and M. Milano. Global cut framework for removing symmetries. In *CP2001*, pages 93–108. Springer, November 2001.
6. E.C. Freuder. Eliminating interchangeable values in constraint satisfaction problems. In *AAAI 91*, pages 227–233, 1991.
7. D. Frost and R. Dechter. Look-ahead value ordering for constraint satisfaction problems. In *Proceedings of the International Joint Conference on Artificial Intelligence, IJCAI'95*, pages 572–578, Montreal, Canada, 1995.
8. R. M. Haralick and G.L. Elliott. Increasing tree search efficiency for constraint satisfaction problems. *Artificial Intelligence*, 14:263–313, 1980.
9. P. Prosser. Hybrid algorithms for the constraint satisfaction problem. *Computational Intelligence*, 9:268–299, 1993.
10. P. Prosser. Binary constraint satisfaction problems: some are harder than others. In *ECAI-94*, pages 95–99, Amsterdam, 1994.
11. J.-F. Puget. Symmetry breaking revisited. In *CP2002*, pages 446–462. Springer, September 2002.

12. T. Schiex and G. Verfaillie. Two approaches to the solution maintenance problem in dynamic constraint satisfaction problems. In *Proc. of the IJCAI-93/SIGMAN Workshop on Knowledge-based Production Planning, Scheduling and Control, Chambery, France, (August 1993).*, 1993.

13. T. Schiex and G. Verfaillie. Nogood Recording for Static and Dynamic Constraint Satisfaction Problem. *International Journal of Artificial Intelligence Tools*, 3(2):187-207, 1994.

14. P. van Beek. Constraint tightness and looseness versus local and global consistency. *Journal of the ACM*, 44(4):549–566, 1997.

15. R. Weigel and B. Faltings. Structuring techniques for constraint satisfaction problems. In *Proceedings of the 15th International Joint Conference on Artificial Intelligence*, pages 418–423, Nagoya, Japan, aug 1997. Morgan-Kaufmann.

Trying Again to Fail-First

J. Christopher Beck[1], Patrick Prosser[2], and Richard J. Wallace[3]

[1] Department of Mechanical & Industrial Engineering, University of Toronto, Canada
jcb@mie.utoronto.ca
[2] Department of Computer Science, University of Glasgow, Scotland
pat@dcs.gla.ac.uk
[3] Cork Constraint Computation Center and Department of Computer Science,
University College Cork, Ireland
r.wallace@4c.ucc.ie

Abstract. For constraint satisfaction problems (CSPs), Haralick & Elliott [1] introduced the Fail-First Principle and defined in it terms of minimizing branch depth. By devising a range of variable ordering heuristics, each in turn trying harder to fail first, Smith & Grant [2] showed that adherence to this strategy does not guarantee reduction in search effort. The present work builds on Smith & Grant. It benefits from the development of a new framework for characterizing heuristic performance that defines two *policies*, one concerned with enhancing the likelihood of correctly extending a partial solution, the other with minimizing the effort to prove insolubility. The Fail-First Principle can be restated as calling for adherence to the second, *fail-first* policy, while discounting the other, *promise* policy. Our work corrects some deficiencies in the work of Smith & Grant, and goes on to confirm their finding that the Fail-First Principle, as originally defined, is insufficient. We then show that adherence to the fail-first policy must be measured in terms of size of insoluble subtrees, not branch depth. We also show that for soluble problems, both policies must be considered in evaluating heuristic performance. Hence, even in its proper form the Fail-First Principle is insufficient. We also show that the "FF" series of heuristics devised by Smith & Grant is a powerful tool for evaluating heuristic performance, including the subtle relations between heuristic features and adherence to a policy.

1 Introduction

Search is at the heart of many AI approaches to problem solving. Despite this importance, there is no understanding at a foundational level of the behavior of heuristic decision making. The answer to the basic question "Why do some heuristics perform better than others?" remains elusive. One long-standing intuition for heuristic performance in Constraint Programming is the Fail-First Principle due to Haralick & Elliott [1] which states: "To succeed, try first where you are most likely to fail." Though initially counter-intuitive, the Fail-First Principle is widely seen as a useful insight into heuristic decision making. Given a set of inter-related decisions, the Fail-First Principle suggests that the one that is most difficult should be made first. This is akin to cautious intelligent behavior, focusing effort on critical choices before allowing ourselves the luxury of solving the easy parts of the problem.

B. Faltings et al. (Eds.): CSCLP 2004, LNAI 3419, pp. 41–55, 2005.

In applying this principle to constraint satisfaction search, Haralick & Elliott made a further critical inference: that minimizing average branch length during search should also minimize search effort. Because of this, their subsequent formal analysis of 'fail-firstness' did not pertain directly to discovering and solving difficult subproblems, but simply to finding the quickest way to fail during search. We will refer to this as the "radical Fail-First Principle" to distinguish it from the original idea. In this analysis, they showed that the "smallest domain first" (SDF) variable ordering heuristic (choose next to assign a value to the variable with the smallest number of possible values) is a level-one estimate of minimizing branch length.

To test this principle, Smith & Grant [2] created a set of new heuristics designed to aggressively fail early in the search. The hypothesis was that if failing quickly was the explanation for search efficiency, then heuristics that failed earlier should demonstrate lower search effort. Smith & Grant found that, contrary to expectations, increasing the ability to fail early in the search did not always lead to increased search efficiency. They concluded that the (radical) Fail-First Principle cannot be the only thing that explains differences in search cost among heuristics.

Our interest in this problem was sparked by the recent development of a new framework for analyzing heuristic performance [3, 4]. This framework incorporates *fail-firstness* as one of its performance principles. Another, called *promise*, concerns the selection of alternatives most likely to succeed. (This principle should be distinguished from the heuristics of the same name [5]; obviously, however, promise *heuristics* are designed to conform to the promise policy if and when the latter applies.) In this framework, if search deviates from a correct path, then and only then, does fail-firstness come into play.

In this paper, we review the work of Smith & Grant and then provide an overview of our policy-based framework for understanding search heuristics. We discuss work which corrects the experiments performed by Smith & Grant; although our results differ in some respects, like the earlier authors we find discrepancies between expectations based on the radical Fail-First Principle and relative performance of their heuristics. We also find that when forward checking is replaced with maintaining arc-consistency (MAC), the radical Fail-First Principle is *supported*. We then show that if fail-firstness is measured using the mean size of insoluble subtrees, adherence to promise and fail-firstness together can account for the behavior of heuristics when either forward checking or MAC is used.

2   Trying Harder to Fail First (1998)

The basic hypothesis of Smith & Grant was that if failing first is such a good thing then more of it must be better. Therefore, creating heuristics with a stronger ability to fail early should increase search efficiency. Efficiency was defined as the number of constraint checks required to find a solution to a problem or to prove that no solution exists. Because the focus of the study was on the relation between fail-firstness and search efficiency, the computational effort to make those heuristic decisions was (correctly) factored out of the experiments. The goal was to understand the relationship between the radical Fail-First Principle and search efficiency so the computational effort to increase the fail-firstness of a heuristic was irrelevant.

Experiments were performed over randomly generated binary constraint satisfaction problems. Each set of problems was defined by a 4-tuple $\langle n, m, p_1, p_2 \rangle$, where n is the number of variables, m is the uniform domain size, p_1 is the proportion of edges in the constraint graph, and p_2 is the uniform constraint tightness. All experiments were over problems with $n = 20$ and $m = 10$.

Using the forward checking algorithm [1] and standard chronological backtracking Smith & Grant tested four heuristics engineered for increasing levels of fail-firstness.

- **FF:** FF is the same as SDF (Smallest Domain First): choose the variable with the smallest remaining domain.
- **FF2:** The variable, v_i, chosen is the one that maximizes $(1 - (1 - p_2^m)^{d_i})^{m_i}$, where m_i is the current domain size of v_i, and d_i is the future degree of v_i. The FF2 heuristic takes into account an estimate (based on the initial parameters of problem generation) of the extent to which each value of v_i is likely to be consistent with the future variables of v_i. The FF2 heuristic has the flavor of the Brelaz heuristic [6], in that it trades off domain size and forward degree and favors differences in the former over the latter.
- **FF3:** FF3 builds on FF2 by using the current domain size of future variables rather than m. The variable, v_i, chosen is the one that maximizes the expression (1) below, where C is the set of all constraints in the problem, F is the set of unassigned variables, and $P = p_2$.
- **FF4:** Finally, FF4 modifies FF3 by using the current tightness, $P = p_{ij}$, of the future constraints (the fraction of tuples from the cross-product of the current domains that fail to satisfy the constraint) instead of p_2.

$$(1 - \prod_{(v_i,v_j) \in C, v_j \in F} (1 - P^{m_j}))^{m_i} \tag{1}$$

The progression from FF through to FF4 uses more and more information in determining which variable is most likely to fail at any given stage of search. Smith & Grant tested this by measuring the distribution of the depth of backtracks for problem instances at $\langle 20, 10, 0.5, 0.37 \rangle$. Their measurements showed that heuristics performed as expected: FF4 has a distribution skewed to backtracks at a shallower depth in search while the distribution gradually moved to deeper backtracks for FF3, FF2, and FF.

With respect to total search effort, Smith & Grant expected that the heuristics would be ranked as follows: FF > FF2 > FF3 > FF4, where > means "results in greater search effort than". Their results were not as expected. Through a number of experiments Smith & Grant showed that except on easy problems, FF2 incurred the least search effort, followed by FF3, FF and finally FF4. That is, they observed the following order: FF4 > FF > FF3 > FF2. This ordering is clearly at odds with the hypothesis of a simple mapping between the ability to fail-first and search effort. They concluded that there must be some other factor at work, perhaps in concert with the Fail-First Principle, in determining search efficiency for variable ordering heuristics.

3   A Framework for Understanding Search

3.1   Policies and Heuristics

The present work is informed by a recently developed framework for characterizing the performance of search heuristics. This framework has two primary elements. A *policy* identifies goals or end-results that are desirable. A *heuristic* is a rule that is followed to make a decision.

For search problems, there is an overall policy of minimizing search effort. This is normally measured by counting nodes or constraint checks in the search tree. Of greater interest is that two subordinate policies can be distinguished in the search domain. When search is in a state that has solutions in its subtree, search effort will be minimized by making decisions to remain on a path to a solution. As this suggests making decisions to move to the most promising subtree, we call this the *promise policy*. However, for hard problems the best choice will not be made in all cases and search may enter a state where the subtree below it does not contain any solutions. In this case, to minimize effort search should fail as quickly as possible so it can return to a path that leads to a solution. We call this the *fail-first policy*.

Heuristics are based on features of the situation that serve to distinguish choices, so that a selection in these terms increases the likelihood of achieving a goal. In CSP search, these are the variable and value ordering "rules" that exist in the constraint literature (e.g.smallest domain first, brelaz [6], domdeg [7]). The intuition behind variable ordering heuristics is the Fail-First Principle mentioned above, while that for value ordering is related to promise. However, recent work, which has shown how to evaluate variable ordering heuristics in terms of promise, indicates that this policy must also be taken into account in any full evaluation of these heuristics [3, 4].

The contribution of heuristic decisions to performance should depend on how well the heuristic conforms to either subordinate policy. We would expect that adherence to the promise policy will make a difference to search for problems with many solutions. As problems become more difficult, the proportion of time exploring bad subtrees becomes greater, so that the fail-first policy is more often in force and fidelity to that policy should be more important. If problems have no solutions, then the only policy relevant to search effort is fail-first.

3.2   The Fail-First Principle

Within the policy framework the Fail-First Principle states that when we are uncertain as to which policy to adhere to (i.e. we do not know if the current search node is good or bad), we should try to adhere to the fail-first policy. Stated in these terms, the Fail-First Principle is a kind of high-level heuristic for selecting a policy to adhere to under conditions of ignorance, where one does not know what the appropriate policy actually is. As such, it is a conservative mini-max principle that tries to minimize worst case effort by aggressively seeking to fail. Adherence to this principle also implies that heuristics should be evaluated in terms of how well they conform to the fail-first policy alone.

With this restatement of the principle, we can see more clearly that there may be important limits to its range of application, and that for certain problems or conditions of search it may lead us badly astray. Moreover, there is the assumption in all of this that the promise policy is irrelevant to variable selection. As we have shown, this is not so: an adequate evaluation of ordering heuristics must give some consideration to both policies rather than the fail-first policy alone. We have also shown, perhaps surprisingly, that heuristics such as SDF which were designed to have a high degree of fail-firstness [1] also show a high degree of promise. This means that for problems with solutions, a correlation between fail-firstness and decreased search effort is not sufficient evidence that this factor is critical for differences in performance, since the impact of this factor has not been disentangled from that of promise.

3.3   Measuring the Ability to Fail-First as Branch Depth Minimization

Within this framework we can also restate the Smith & Grant strategy: by devising heuristics to conform in increasing degree to the fail-first policy, we can evaluate the sufficiency of the Fail-First Principle. However, to evaluate sufficiency we must be able to measure fail-firstness, i.e. the degree of adherence to the policy. For now we will continue with the assumption made by Haralick & Elliott and by Smith & Grant that fail-firstness is adequately characterized by branch depth.

For their evaluation, Smith & Grant measured the mean *backtrack depth* to find a solution to a problem or to prove that no solution exists. We believe there are two weaknesses in this methodology. First, a backtrack is counted whenever a domain is emptied and search returns to the previous variable. If that variable has no more values to try, its domain has also been emptied and another backtrack is counted in moving back once again. That is, Smith & Grant measured the average depth of failed leaf nodes *and* failed interior nodes of the search tree explored. The original formulation of the radical Fail-First Principle assumed that minimizing mean *branch depth* would minimize search effort. Therefore, it should be the mean branch depth that is used as a measure of the ability of a heuristic to fail-first. The backtrack depth measure does not do this. A more appropriate measure of the mean branch depth is as follows: whenever a variable is assigned a value and that assignment immediately leads to a domain wipe-out, we count a failure. That is, we measure the average depth of failed leaf nodes in the search tree.

There is an alternative way of measuring branch depth: calculate the difference between the depth of a failed leaf and the depth of the initial mistake that led to the failure. Although this seems like an even more precise measure, it suffers from the effect of a varying ceiling: the largest possible difference is greater when mistakes occur higher in the search tree. So this measure was not used for the initial experiments, although data will be presented in Section 5.

The second weakness with the earlier measurement of a heuristic's ability to fail-first is that by searching only for the first solution to soluble problems the measure of the ability to escape a mistake is contaminated by the heuristic's ability to find a solution (its promise, in the newer formulation). It is at least theoretically possible that a heuristic that is very poor at failing first could be very good at finding a solution when one exists. Therefore, searching for the first solution combines the abilities of a heuristic to escape dead-ends and its ability to find solutions. We are interested in isolating the former ability. For our experiments, we assess the ability of a heuristic to fail-first by

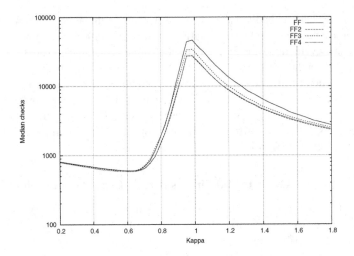

Fig. 1. The median number of checks for the $\langle 20, 10, 0.5 \rangle$ problem set. The heuristics are ranked as follows: FF > FF3 > FF4 $\approx$ FF2.

measuring the branch depth in searching for *all* solutions or showing that no solution exists. (Of course, performance assessment is still based on the work required to find the first solution.)

4 Empirical Investigations

In this section, we report the results of three experiments. The first two explore the relationship between fail-firstness and heuristic quality. They were reported in part by Beck et al. [8], and are included here (with further data) for completeness. The conclusion of the first experiment is that, as Smith & Grant originally found, fail-firstness as measured by branch depth is not equivalent to heuristic quality. The third experiment, therefore, investigates whether promise is the missing factor.

4.1 Fail-First with Forward Checking

As reported earlier [8], we repeated the experiments of Smith & Grant. Our results were produced by using two solvers coded independently, and we also confirmed these results using the C++ solver of Smith & Grant with an error in FF4 corrected[4]. In our implementation of the heuristics, ties (when more than one variable is judged heuristically best) are broken lexicographically. The problems we investigated were generated at the beginning of our study and stored. They were then used by all our solvers, allowing us to reproduce results across two sites.

Problems were generated using a "probability-of-inclusion" model, in which each possible constraint element (domain value, constraint or constraint tuple) is included with a specified probability. The generator allows parameters to be fixed at the expected

---

[4] The value of p_{ij} was computed incorrectly for the FF4 heuristic in [2].

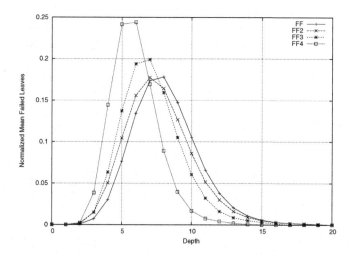

Fig. 2. The fraction of failed leaves at each depth in the tree when searching for all solutions for 1000 problems in the $\langle 20, 10, 0.5, 0.37 \rangle$ set, i.e. the problems at the $\langle 20, 10, 0.5 \rangle$ phase transition.

values given these probability values; in this case a set of elements is generated repeatedly until the cardinality matches the expected value. This allows it to generate problems in accordance with model B [9]. By specifying a probability of 1 for domain element inclusion, all domains have a size specified by a maximum domain-size parameter. Sets of soluble or insoluble problems were sometimes used; these problems were generated in an identical fashion and filtered on the basis of solution testing.

Figure 1 presents the median number of constraint checks to find a solution or to prove that no solution exists for problems from the set $\langle 20, 10, 0.5 \rangle$. The constraint tightness was varied from 0.01 to 0.99 in steps of 0.01. For each combination of parameters 1000 problem instances were generated. Median checks (the same measure used by Smith & Grant) are plotted against κ (kappa), the measure of constrainedness proposed by [10]. Similar results were found for other problem series [8].

In these experiments, FF clearly incurs the highest number of constraint checks, followed by FF3. In Figure 1, there is no discernible difference between FF2 and FF4. Aside from the FF4 results, these graphs agree with the original experiments of Smith & Grant.

In Figure 2 we plot the fraction of failed leaves at each depth when searching for all solutions. Qualitatively, these results match what Smith & Grant found with their measure of the ability to fail-first: FF4 does indeed fail higher in the tree (as judged by mean branch depth), followed by FF3, FF2 and FF. Therefore, we can confirm that Smith & Grant did indeed propose new heuristics that progressively increase in their ability to minimize branch depth.

The crux of these experiments is the fact that the ability to fail earlier in the tree does not necessarily translate into better search effort; FF3 incurs a higher search cost than both FF2 and FF4, yet Figure 2 shows that FF3 is between FF2 and FF4 in its ability to fail early in the tree. These results have been extended to larger problems, where the ordering with respect to search effort is clearly FF > FF3 > FF2 > FF4 [8].

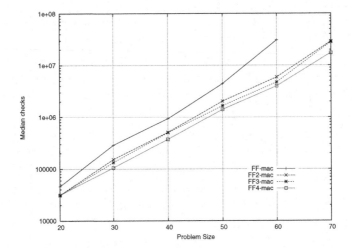

Fig. 3. Median consistency checks for problems with 20 to 70 variables using MAC. All problems have 10 values per variable, a density that results in 10 constraints on each variable, and a tightness set to give $\kappa \approx 0.9$. All problems are soluble and each problem size has 50 instances.

Thus, even after fixing the errors and improving the measurement of fail-firstness in terms of branch depth, the results do not entirely conform to expectations based on the Fail-First Principle as it was originally formulated. This confirms the main conclusion of Smith & Grant: the radical Fail-First Principle (i.e., branch depth minimization) is not sufficient to account for all variations in search efficiency among variable ordering heuristics.

4.2   Tests with MAC

In the course of this work we also wanted to determine whether the consistency enforcement algorithm might have an effect on the heuristics' adherence to the fail-first policy, and hence the viability of the Fail-First Principle under different degrees of consistency maintenance. To test this we repeated the above experiments, varying problem size, but this time using the maintaining-arc consistency algorithm (MAC) [11]. As the name implies, whenever a variable is instantiated the future sub-problem is made arc-consistent. If this results in a domain wipe-out, a new value is tried, and failing that, backtracking takes place.

For 20-variable problems, there is little variation in fail-depth among heuristics, although the differences that appear at different depths are consistent with the pattern found for forward checking. For this reason we present data on larger problems where a clear pattern of differences emerges.

The results for search efficiency are presented in Figure 3. Interestingly, the ranking of heuristics is different than for FC. We now have the order FF > FF2 > FF3 > FF4. The results in Figure 3 are for soluble problems, and though not shown, the same ranking was found for insoluble problems. Figure 4 shows the failure depths of the

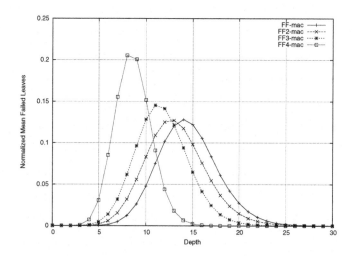

Fig. 4. Distribution of branch depth for soluble problems with 60 variables (parameters $\langle 60, 10, 0.153, 0.369 \rangle$). Qualitatively similar results are seen for insoluble problems.

heuristics when allied with the MAC algorithm, for one set of large problems ($n = 60$). The results are in agreement with those in Figure 2 with FF4 failing earliest, then FF3, FF2, and FF.

Thus, we find that for MAC, differences among heuristics are entirely in line with expectations based on the radical Fail-First Principle. Adherence to the principle appears to be algorithm dependent.

4.3   The Impact of Promise

We have already noted that the efficiency of variable ordering heuristics can be related to promise as well as fail-firstness. It was therefore of interest to see whether FF heuristics could be distinguished on this basis. Figure 5 shows the results of "probe" tests, which can be used to provide unbiased estimates of promise [3]. We find that with forward checking FF3 and FF4 are distinctly inferior to FF and FF2 on this basis.

However, before we conclude that differences in promise are the whole explanation for the order of search effort, we must consider insoluble problems. As noted earlier, promise is only well-defined on problem instances with solutions. If the results on search effort are due to a combination of promise *and* fail-firstness, then the relative ranking of heuristics should differ for soluble and insoluble problems. We therefore refine Smith & Grant's hypothesis and instead hypothesize that on *insoluble* problems as we try harder to minimize branch depth we will reduce search effort.

Figure 6 demonstrates that this hypothesis is false. The relative ordering of the heuristics on insoluble problems for the $\langle 20, 10.0.5 \rangle$ problem set is identical to the ordering on mixed problems. Though not shown, this ordering is the same for the soluble problems as well. Furthermore, the branch depth on insoluble (as well as soluble) problems follows the pattern of Figure 2: FF4 fails highest followed by FF3, FF2, and FF.

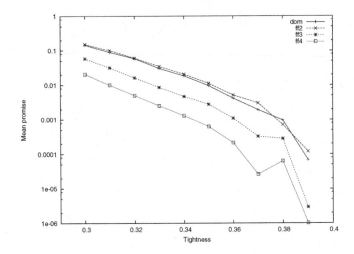

Fig. 5. Promise measurements for the FF heuristics with forward checking, obtained for soluble <20,10,0.5> problems.

Therefore promise cannot be the entire explanation for the discrepancies in search effort. (As we will see shortly, it would be premature to conclude that it does not play a role in the case of soluble problems.)

4.4   Conclusions

Three conclusions are drawn from the experiments above:

1. For forward checking, minimizing mean branch depth does not wholly account for heuristic efficiency.
2. For forward checking, minimizing mean branch depth combined with adherence to the promise policy does not wholly account for heuristic efficiency.
3. For MAC, minimizing mean branch depth does appear to account for heuristic efficiency.

A model of heuristic performance needs to explain each of these results. From the perspective of our policy framework, there appear to be two logical options. Either there is an additional policy that accounts for the discrepancies or the measurements that we have proposed for promise and fail-firstness are deficient. For the balance of this paper, we examine the second option, concentrating on the fail-firstness measurement.

5   The Proper Measure of Fail-Firstness

The fail-first policy is to minimize the effort required to determine that an insoluble subtree is indeed insoluble. This suggests that the average size of insoluble trees would give a more accurate measure of fail-firstness than measures of branch depth. Something like this was suggested by Nudel [12], who argued that the minimum domain size (FF)

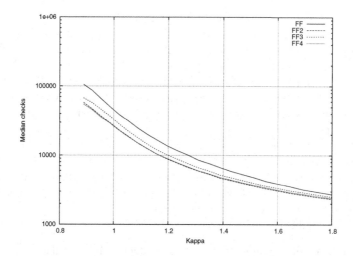

Fig. 6. The median number of constraint checks for the insoluble problem instances in the ⟨20, 10, 0.5⟩ problem set. The problem sets plotted are those for which at least one instance is insoluble, therefore each κ value may have a different number of instances up to 1000. In particular, the low values of κ are based on few instances.

heuristic worked by minimizing the size of failed subtrees. Failed subtree size takes into account the branching factor in the search tree. As Smith & Grant point out, Russell & Norvig [13] suggested that successful heuristics such as Brelaz and FF can be justified by the minimization of the branching factor. However, the fail-first policy suggests that a combination of this and branch depth may be critical, and this is more likely if there is a tradeoff between the two.

To examine failed subtree size, a more detailed assessment of search characteristics was performed using the 50-variable problems from the series tested earlier ([8] and Section 4.2; their parameters were ⟨50, 10, 0.184, 0.369⟩). (Similar results were also obtained in tests on 20-variable problems of varying tightness.) In addition to mean fail-depth, we calculated the following measures:

1. Mean fail-length: the difference in search depth between the initial invalid assignment, or "mistake" and the detection of failure
2. Mean "mistake-tree size" (mistksz): the number of nodes in an insoluble subtree, rooted at the initial invalid assignment
3. Number of failures
4. Number of mistake-trees
5. Mean selected domain size ($|\,d\,|$): the domain size of the variable which is selected for assignment. This is a direct measure of the branching factor.

The basic summary results are shown in Table 1 for each heuristic for forward checking and MAC. These results must be considered in tandem with similar measures that take the depth of the initial mistake into account. This is because of the problem mentioned earlier, that intensity measures like fail-length and mistake-tree size involve ceiling effects related to the location of a mistake in the search tree. For example, if a mistake is made at level n-1, then the maximum size of the mistake-tree is 1; this means that if, for a given heuristic a large proportion of the mistakes occur at this level the average mistake-tree size will be small because of the low ceiling in these cases. Table 2 displays the fail-length and mistake-tree size at different mistake depths.

In both tables and for each subcolumn the statistical significance of differences between entries was assessed by ANOVAs and post-hoc comparisons between pairs of means that were closest in value [14]. Analyses for Table 2 were based on problems for which there were no zero frequencies for any heuristic. (Because of the number of zero frequencies, analyses were not carried out beyond level 20 for FC and level 10 for MAC.) For all ANOVAs, the F statistic was statistically significant at $p < .05$ or better (typically $\leq 10^{-8}$). The few comparisons between adjacent means that were not statistically significant are indicated in the tables.

There are a number of important results in these tables.

- The ordering of fail-lengths matches that of fail-depths (Table 1), and the same ordering of fail-lengths is found at every mistake depth (Table 2). Since neither measure of branch depth corresponds to the order of search effort, this supports the conclusion that measures of branch depth are inadequate measures of fail-firstness.
- Further evidence on this point comes from a comparison of fail-length and mistake-tree size at mistake depth 1. These results are directly comparable to those with insoluble problems, since in the latter case all mistakes occur at this level. (This has been verified directly on insoluble problems.) In these cases fail-first is the only relevant policy. Here we see that the order of mistake-tree size matches the order of search effort (cf. Figure 6), while the fail-length order does not. This implies that any measure of fail-firstness must be based on the size of insoluble subtrees.
- For these soluble problems, FF3 with forward checking produces more mistake-trees (and consequent failures) than any other heuristic (Table 1). This reflects a relative failure to adhere to the promise policy, so poor performance of FF3 in this case must be partly due to this factor.
- With forward checking, differences in mistake-tree size vary as a function of mistake depth (Table 2); as depth increases, FF3 becomes better than FF2. This shows that relative adherence to a policy by different heuristics can vary at different levels of search.

We conclude that the difference from expectation found for FF3 in the first three experiments is related to deficiencies with respect to both policies. For soluble problems, the fail-first deficiency occurs at the top of the search tree, but there is also a deficiency in promise, reflected in the number of mistakes made by this heuristic. For insoluble problems, since fail-first is the only relevant policy, the deficiency is restricted to this factor. A more general conclusion is that when fail-firstness is properly measured, the order of search effort obtained by these heuristics can be characterized by degree of adherence to one or both policies. Finally, we have shown that valid comparisons of fail-firstness must be made at the same mistake depth. In fact, there are important differences among heuristics with respect to fail-firstness that are a function of mistake depth (as well as algorithm type).

Table 1. Failure measures and branching factor for fail-first heuristics.

	faildepth	fail-length	#fails	mistksz	#mstktrs	| d |
		forward checking				
FF	19.9	17.5	600,303	6679	1662	1.35
FF2	18.3	15.9	103,365	1799	539	1.36
FF3	16.6	13.8	1,030,002	644	10,914	2.25
FF4	13.7	10.9	289,466	339	4885	2.18
		MAC				
FF	12.3	10.0	59,410	786	324	2.09
FF2	11.3	8.9	13,662	247	175	1.64
FF3	10.1	7.8	14,056	203	235	1.37
FF4	7.6	5.4	9,401	119	193	1.25

Note. Values are means per problem. 50-variable problems with solutions. "#fails" is number of times an assignment led to a domain wipeout. "mistksz" is mistake-tree size; this and other measures are defined in text. "| d |" is mean domain size of variables selected for instantiation. In this and Table 2 nonsignificant comparisons are flagged by underlining the smaller value.

Table 1 suggests that the branching factor plays a significant role in the relative efficiency of these heuristics. With forward checking, FF3 and FF4 each had a higher branching factor than the other heuristics, and it was here that inversions in the expected ordering occurred. (In addition to the means shown in the table, the *range* of domain sizes of the variables selected was much greater with FF3 and FF4, so that sizes up to 10 were chosen even at moderate depths of the search tree.) With MAC, in contrast, the branching factor diminished from FF through FF4, and in this case search effort followed the original expectations. We hypothesize that, with forward checking, the greater branching of FF3 overwhelms the modest improvement in branch depth, which leads to inferior search performance in comparison with FF2.

That it is a tradeoff between branch depth and the branching factor can be seen by comparing the FF and FF4 results for forward checking. If degree of branching were the only factor, FF should incur less search effort than FF4. As shown by all our experiments, this is not the case.

6   Conclusions

This work serves both to clarify the Fail-First Principle and to indicate its limitations. This has been done by using a new framework in which heuristics are evaluated in terms of how well they adhere to either of two basic policies, called promise and fail-first, one of which is in force at any point in search.

In order to evaluate variable ordering heuristics in these terms we have developed measures of adherence to a given policy. For promise, this was done in earlier work [3, 4]. In the present paper, we have done this for fail-firstness. An important contribution of this work is to demonstrate that fail-firstness cannot be adequately assessed by measures of branch depth; instead the size of the insoluble subtree rooted at the initial mistake must be calculated.

Table 2. Failure measures for fail-first heuristics at different depths of search tree.

	mistake depth (nodes)					
	1	5	10	15	20	30
			FC			
			fail-length			
FF	18.2	15.5	11.8	9.6	7.7	4.0
FF2	16.4	14.2	11.2	8.3	5.9	2.1
FF3	14.4	11.8	9.3	6.2	4.5	0.8
FF4	11.6	8.8	5.8	2.7	1.2	0.0
			mistake-tree size			
FF	145,015	4666	219	52	21	10
FF2	24,808	1549	150	27	12	3
FF3	80,484	3154	188	22	8	2
FF4	21,813	678	35	5	2	1
			MAC			
			fail-length			
FF	10.6	7.9	3.9	1.7	0.5	0.0
FF2	9.6	6.6	3.3	1.7	0.9	0.0
FF3	8.3	5.5	2.2	0.7	0.2	0.0
FF4	6.0	3.2	0.4	0.1	0.0	–
			mistake-tree size			
FF	7080	219	12	3	1	1
FF2	1903	94	10	3	2	1
FF3	1489	63	6	2	1	1
FF4	865	25	2	1	1	–

Note. Values are means per problem for each mistake depth. "Mistake depth" is depth at which an initial assignment was made to produce an insoluble subtree.

Within this framework, we can restate the Fail-First Principle as a metaheuristic proposal to act as if adherence to the fail-first policy is the only important consideration. Having cleared up the question of how to measure such adherence, the original conclusion of Smith & Grant, that the radical Fail-First Principle that was proposed by Haralick & Elliott is inadequate, is of course confirmed, since this form of the principle is based on an incorrect measure of fail-firstness. At the same time, we have shown that Smith & Grant's assumption that their fail-first series of heuristics shows increasing adherence to this policy is not entirely correct, since with forward checking average mistake-tree size is sometimes greater for FF3 than for FF2, and is always greater for insoluble problems. Finally, we have shown that even when put in its proper form, the Fail-First Principle is not an entirely reliable guide (cf. results in Table 1, showing that average mistake-tree size for FF3 is less than FF2); this is because variation in adherence to the promise policy can be more important than variation in adherence to the fail-first policy.

Clearly, features such as the branching factor determine whether a heuristic adheres to either policy or to both. What does the policy framework give us in addition? We think the answer is that it gives us a way of characterizing the effects of heuristic features on performance in terms of basic features of search itself. In addition, the present work shows that we can evaluate performance in these terms without knowing the features of the heuristic that affect its adherence to a policy. Finally, we know from the present work as well as the original study of Smith & Grant that there is no simple relation between heuristic features and adherence to a policy, and there is no guarantee that improved adherence to one policy will result in improved adherence to the other. These seem to be sufficient reasons for making the distinctions that the new framework requires.

Acknowledgment

This work was supported by Science Foundation Ireland under Grant 00/PI.1/C075 and ILOG S.A. We thank Diego Moya for some of the coding and especially Barbara Smith for giving us access to source code and supporting our study.

References

1. Haralick, R.M., Elliott, G.L.: Increasing tree search efficiency for constraint satisfaction problems. Artificial Intelligence **14** (1980) 263–314
2. Smith, B.M., Grant, S.A.: Trying harder to fail first. In: Proc. Thirteenth European Conference on Artificial Intelligence-ECAI'98, John Wiley & Sons (1998) 249–253
3. Beck, J.C., Prosser, P., Wallace, R.J.: Toward understanding variable ordering heuristics for constraint satisfaction problems. In: Proc. Fourteenth Irish Artificial Intelligence and Cognitive Science Conference-AICS'03. (2003) 11–16
4. Beck, J.C., Prosser, P., Wallace, R.J.: Variable ordering heuristics show promise. In: Principles and Practice of Constraint Programming-CP'04. LNCS No. 3258. (2004) 711–715
5. Geelen, P.A.: Dual viewpoint heuristics for binary constraint satisfaction problems. In: Proc. Tenth European Conference on Artificial Intelligence-ECAI'92. (1992) 31–35
6. Brelaz, D.: New Methods to Color the Vertices of a Graph. Communications of the ACM **22** (1979) 251–256
7. Gent, I., MacIntyre, E., Prosser, P., Smith, B., Walsh, T.: An empirical study of dynamic variable ordering heuristics for the constraint satisfaction problem. In: Principles and Practice of Constraint Programming-CP'96. LNCS No. 1118. (1996) 179–193
8. Beck, J.C., Prosser, P., Wallace, R.J.: Failing first: An update. In: Proc. Sixteenth European Conference on Artificial Intelligence-ECAI'04. (2004) 959–960
9. Gent, I.P., MacIntyre, E., Prosser, P., Smith, B.M., Walsh, T.: Random constraint satisfaction: Flaws and structure. Constraints **6** (2001) 345–372
10. Gent, I.P., MacIntyre, E., Prosser, P., Walsh, T.: The constrainedness of search. In: Proc. Thirteenth National Conference on Artificial Intelligence-AAAI'96. (1996) 246–252
11. Sabin, D., Freuder, E.: Contradicting Conventional Wisdom in Constraint Satisfaction. In: Proc. Eleventh European Conference on Artificial Intelligence-ECAI'94, John Wiley & Sons (1994) 125–129
12. Nudel, B.: Consistent-labeling problems and their algorithms: Expected-complexities and theory-based heuristics. Artificial Intelligence **21** (1983) 263–313
13. Russell, S.J., Norvig, P.: Artificial Intelligence: A Modern Approach. Prentice-Hall (1995)
14. Hays, W.L.: Statistics for the Social Sciences. 2 edn. Holt, Rinehart, Winston (1973)

Characterization of a New Restart Strategy
for Randomized Backtrack Search

Venkata Praveen Guddeti and Berthe Y. Choueiry

Constraint Systems Laboratory,
Computer Science & Engineering,
University of Nebraska-Lincoln
{vguddeti,choueiry}@cse.unl.edu

Abstract. We propose an improved restart strategy for randomized backtrack search, and evaluate its performance by comparing to other heuristic and stochastic search techniques for solving random problems and a tight real-world resource allocation problem. The restart strategy proposed by Gomes et al. [1] requires the specification of a cutoff value determined from an overall profile of the cost of search for solving the problem. When no such profile is known, the cutoff value is chosen by trial-and-error. The Randomization and Geometric Restart (RGR) proposed by Walsh does not rely on a cost profile but determines the cutoff value as a function of a constant parameter and the number of variables in the problem [2]. Unlike these strategies, which have fixed restart schedules, our technique (RDGR) dynamically adapts the value of the cutoff parameter to the results of the search process. Our experiments investigate the behavior of these techniques using the cumulative distribution of the solutions, over different run-time durations, values of the cutoff, and problem types. We show that distinguishing between solvable and over-constrained problem instances yields new insights on the relative performance of the search techniques tested. We propose to use this characterization as a basis for building new strategies of cooperative, hybrid search.

1   Introduction

We have developed a system for modeling and solving a resource allocation problem, which is the assignment of Graduate Teaching Assistants (GTA) to courses in our department [3]. We exploit this system as a platform for developing and characterizing new problem-solving strategies. The research we describe in this paper was motivated and enabled by this project. However, our results are here extended beyond this particular application.

The Graduate Teaching Assistants Assignment Problem (GTAAP) is a critical and arduous task that the department's administration has to drudge through every semester. By focusing our investigations on this particular real-world application, we have been able to identify and compare the advantages and shortcomings of the various search strategies we have implemented to solve this problem. Such an insight is unlikely to be gained from testing toy problems, and surely difficult from testing random problems. We show that the identified behaviors apply beyond our application. The contributions of this paper are as follows: (1) The development of a new dynamic restart strategy for

B. Faltings et al. (Eds.): CSCLP 2004, LNAI 3419, pp. 56–70, 2005.

randomized backtrack search, and (2) an empirical evaluation of the performance of this new strategy and a comparison with other heuristic and stochastic search techniques on a real-world problem and on randomly generated binary CSPs.

This paper is structured as follows. Section 2 describes the GTA assignment problem (GTAAP) and our implementations of a backtrack search, a local search, and a multi-agent search technique for solving it. Section 3 introduces our new proposed dynamic restart strategy for randomized backtrack search and our implementation of Walsh's restart strategy [2]. Section 4 presents our experiments and our observations. Finally, Section 5 concludes the paper and provides directions for future research.

2   GTA Assignment Problem

Given a set of Graduate Teaching Assistants (GTAs), a set of courses, and a set of constraints that specify allowable assignments of GTAs to courses, the goal is to find a consistent and satisfactory assignment [4–6]. Hard constraints (e.g., a GTA's competence, availability, and employment capacity) must be met, and GTA's preferences for courses (expressed on a scale from 0 to 5) must be maximized. Typically, every semester, the department has about 70 different academic tasks and can hire between 25 and 40 GTAs. Instances of this problem, collected since Spring 2001, are consistently tight and often over-constrained. The objective is to ensure GTA support to as many courses as possible by finding a *maximal consistent partial-assignment*. Because the hard constraints cannot be violated, the problem cannot be modeled as a MAX-CSP [7]. Our constraint-model represents the courses as variables, the GTAs as domain values, and the assignment rules as a number of unary, binary, and non-binary constraints. We define the problem as the task of finding the longest assignment, as a primary criterion, and maximizing GTAs' preferences, as a secondary criterion. (We model the latter as the value of the geometric mean of GTAs' preferences in an assignment.) We implemented a number of search strategies for solving this problem (summarized below). We tested these search techniques on the real-world data-sets shown in Table 1. Each course has a load that indicates the weight of the course. For example, a value of 0.5 means this course needs one-half of a GTA. The *total load* of a semester is the cumulative load of the individual courses. Each GTA has a capacity factor which indicates the maximum course weight he/she can be assigned during the semester. The sum of the capacities of all GTAs represents the *total capacity*.

Table 1. Characteristics of the data sets.

Data set	Spring2001b	Fall2001b	Fall2002	Fall2002-NP	Spring2003	Spring2003-NP
Reference	1	2	3	4	5	6
Solvable?	×	✓	×	×	✓	✓
#Variables	69	65	31	59	54	64
Max domain size	26	34	28	28	34	34
Total capacity	26	30	11.5	27	27.5	31
Total load	29.6	29.3	13	29.5	27.4	30.2
Ratio = $\frac{\text{Total Capacity}}{\text{Total Load}}$	0.88	1.02	0.88	0.91	1.00	1.02

We compare our new dynamic restart strategy (RDGR) with a heuristic backtrack search (BT) with various ordering heuristics, a greedy local search (LS), a multi-agent-based search (ERA), and a randomized backtrack search with restart (RGR). All strategies implement the above two optimization criteria, except ERA, which models the GTAAP as a satisfaction problem. These search techniques were separately implemented on the same model and data structures by students competing to produce the best results. Below, we summarize the design of BT, LS, and ERA.

2.1 Heuristic Backtrack Search

Our heuristic backtrack (BT) search is a depth-first search with forward checking [8]. Because the problem may be over-constrained, we modified the backtrack mechanism to allow null assignments and proceed toward the longest solution in a branch-and-bound manner (i.e., backtracking is not performed when a domain is wiped-out as long as there are future variables with no empty domains). This solution is equivalent to adding a dummy value in the variables' domains. Our implementation is described in detail in [5]. We have implemented several ordering heuristics to improve the performance of search (see [9]). Our experiments showed that dynamic variable ordering is consistently superior to static ordering, but that the influence of the other factors is not significant in the context of our application.

All these strategies exhibited a serious vulnerability to thrashing (i.e., searching unpromising parts of the search space), which seriously undermined their ability to explore wider areas of the search space. Indeed, although BT is theoretically sound and complete, *the size of the search space makes such guarantees meaningless in practice.* Figure 1 illustrates the gravity of thrashing for a problem with 69 variables and 26 values. We define the 'shallowest level' as the shallowest level in a search tree attained by the backtracking mechanism along any given path. The percentage denotes the ratio $\frac{\text{number of variables} - \text{shallowest level}}{\text{number of variables}}$. Indeed, the shallowest level of backtrack achieved after 24 hours (26%) is not significantly better than that reached after 1 minute (20%) of search, never revising the initial assignment of 74% of the variables. Figure 2 shows, for each data set, the number of variables, the longest solution (max depth), and the shallowest BT levels in terms of the level and the percentage of backtracking in the search tree attained after 5 minutes and 6 hours.

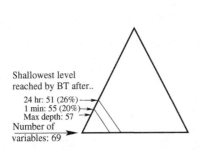

Shallowest level
reached by BT after..

24 hr: 51 (26%)
1 min: 55 (20%)
Max depth: 57
Number of
variables: 69

Data	#	BT running for..					
set	Vars	5 min			6 hours		
		Max depth	Shallowest level	%	Max depth	Shallowest level	%
1	69	57	53	23%	57	51	26 %
2	65	63	55	15%	63	54	16 %
3	31	28	13	58%	28	3	90 %
4	59	49	48	18%	50	45	23 %
5	54	52	44	18%	54	41	24 %
6	64	62	54	15%	62	47	26 %

Fig. 1. BT thrashing in large search spaces.   **Fig. 2.** BT search thrashing.

Table 2. Performance of BT for various running times.

Data set 1 (69 variables, over-constrained)						
Running time	30 sec	5 min	30 min	1 hour	6 hours	24 hours
Shallowest BT level	54	53	52	52	51	51
Longest solution	57	57	57	57	57	57
Geometric mean of preference values	2.15	2.17	2.17	2.21	2.27	2.27
# Backtracks	1835	47951	261536	532787	3274767	13070031
# Nodes visited	3526	89788	486462	989136	6059638	24146133
# Constraint checks	8.50E+07	3.17E+08	1.81E+09	3.58E+09	2.16E+10	8.70E+10

As the problem size increases, the effects of thrashing become more important. Table 2 shows the performance of BT on data set 1 for various run times. Even after letting our best heuristic backtrack search run for over 24 hours, the quality of the solution, in terms of solution length, is not improved. The improvement of the assignment quality, in terms of the geometric mean of the preference values, is insignificant. Finally, we notice that the assignment of the first 51 variables in the ordering was never undone. Consequently, in practice, completeness is a purely theoretical feature.

2.2 Local Search

Zou and Choueiry designed and implemented a greedy, local search (LS) technique for the GTAAP system [10–12]. It is a hill-climbing search using the min-conflict heuristic for value selection [13]. It begins with a complete, random assignment (not necessarily consistent), and tries to improve it by changing inconsistent assignments in order to reduce the number of constraint violations. In order to deal effectively with global constraints (e.g., capacity constraints), we identify, one at a time and in random order, a variable that satisfies all remaining constraints as consistent, and propagate the effects of this consistent assignment by filtering the domains of the remaining variables. This design decision effectively handles non-binary constraints. Our local search is greedy in the sense that consistent assignments are not undone. Moreover, a random-walk strategy is applied to escape from local optima [14]. With a probability $(1 - p)$, the value of a variable is chosen using the min-conflict heuristic, and with probability p this value is chosen randomly. Following the indications of [14] and after testing, $p = 0.02$ is used. Finally, random restarts are used to break out of local optima.

2.3 Multi-agent Search

Liu et al. proposed the ERA algorithm (Environment, Reactive rules, and Agents) as a multi-agent-based search for solving CSPs [15]. Zou and Choueiry implemented and tested ERA for solving GTAAP [10–12]. In ERA, each agent represents a variable. The positions of an agent in the environment correspond to the values in the domain of the variable. Starting from a random positioning of the agents in the environment, each agent evaluates the quality of its positions given the positions of the remaining agents and decides to move to what seems to be the best position, the choice being determined stochastically by the reactive rules. The agents keep moving until they all reach a satisfying position (i.e., a full, consistent solution) or a certain time period has elapsed. This algorithm acts as an 'extremely' decentralized local search, where any agent can move to any position, likely *forcing other agents to seek other positions*. Zou and Choueiry showed that the extreme mobility of agents in the environment is the reason behind ERA's amazing immunity to local optima [10–12]. They found that ERA is indeed the only search technique to solve GTAAP instances that remain unsolved by all other techniques tested. They uncovered the weakness of ERA on over-constrained problems and characterized it as a livelock phenomenon (where some agents keep forcing each others out of chosen positions thus causing cycles and undermining the stability of search). Finally, they showed how this phenomenon can be advantageously used to isolate, identify, and represent conflicts in a compact manner.

3   Randomized BT Search with Restarts

Unlike ERA and local search, general backtrack (BT) search is, in principle, complete and sound. However, the performance of heuristic BT is seriously undermined by thrashing. Thrashing can be explained by incorrect heuristic choices made early in the search process. We explore randomization in BT as a way to overcome this shortcoming of systematic search. First we review the main concepts, then we describe the two strategies that we tested.

Gomes et al. demonstrated that randomization of heuristic choices combined with restart mechanisms is effective in overcoming the effects of thrashing and in reducing the total execution time of systematic BT search [1]. Thrashing in BT search indicates that search is stuck exploring an unpromising part of the search space, and thus incapable of improving the quality of the current solution. It becomes apparent that there is a need to interrupt search and to explore other areas of the search space. It is important to restart search from a different portion of the search space; otherwise it will end up traversing the same paths. Randomization of branching during search is used to this end. Randomness can be introduced in the variable and/or value ordering heuristics, either for tie-breaking or for variable and/or value selection. After choosing a randomization method, the algorithm designer must decide on the type of restart mechanism. This restart mechanism determines when to abandon a particular run and restart the search. Here the tradeoff is that reducing the cutoff time reduces the probability of reaching a solution at a particular run. Several restart strategies have been proposed with different cutoff schedules. Some of the better known ones are the fixed-cutoff strategy and Luby et al.'s universal strategy [16], the randomization and rapid restart (RRR) of Gomes et al. [1], and the randomization and geometric restarts (RGR) of Walsh [2]. Among the above listed restart strategies, RRR and RGR have been studied and empirically tested in the context of CSPs. All of these restart strategies are static in nature, i.e. the cutoff value for each restart is independent of the progress made during search. Some restart strategies (e.g., fixed-cutoff strategy of [16] and RRR [1]) employ an optimal cutoff value that is fixed for *all* the restarts of a particular problem instance. The estimation of the optimal cutoff value requires a priori knowledge of the cost distribution of that problem instance, which is not known in most settings and must be determined by trial-and-error. This is clearly not practical for real-world applications. There are other restart strategies that do not need any a priori knowledge (e.g., Luby et al.'s universal strategy [16] and Walsh's RGR [2]). They utilize the idea of an increasing cutoff value in order to ensure the completeness of search. However, if these restart strategies do not find a solution after the initial few restarts, then the increasing cutoff value leads to fewer restarts, which may yield thrashing and diminishes the benefits of restart. We propose a restart strategy that dynamically adapts the cutoff value for each restart based on the performance of previous restarts. Our strategy loses the guarantee of completeness, which, anyway, is not achievable on large problems.

3.1   Randomization and Geometric Restarts

Walsh proposed the Randomization and Geometric Restarts (RGR) strategy to automate the choice of the cutoff value [2]. According to RGR, search proceeds until it reaches a cutoff value for the number of nodes visited. The cutoff value for each restart is a constant factor, r, larger than the previous run. The initial cutoff is equal to the number of variables n. This fixes the cutoff value of the ith restart at $n.r^i$ nodes. The geometrically increasing cutoff value ensures completeness with the hope of solving the problem before the cutoff value increases to a large value. We studied various values of r and report them in Section 4.2. We combined this restart strategy with the backtrack search of Section 2.1, randomizing the selection of variable-value pairs.

3.2   Randomization and Dynamic Geometric Restarts

We now introduce a simple but effective improvement to RGR. All static restart strategies suffer from the problem of increasing cutoff values after each restart. While this ensures completeness of the search, it results in fewer restarts, thus increasing the likelihood of thrashing and diminishing the probability of finding a solution. Our proposed strategy, Randomization and *Dynamic* Geometric Restarts (RDGR), aims to attenuate this effect. It operates by not increasing the cutoff value for the following restart *whenever the quality of the current best solution is not improved upon.* When the current restart improves on the current best solution, then the cutoff value is increased geometrically, similar to RGR. Because the cutoff value does not necessarily increase, completeness is no longer guaranteed. This situation is acceptable in application domains (like ours) with large problem size where completeness is, anyway, infeasible in practice. Smaller cutoff values result in a larger number of restarts taking place in RDGR than RGR, which increases the probability of finding a solution. All other implementation details are similar to RGR.

Let C_i be cutoff value for the ith restart and r be the ratio used to increase the cutoff value. In RGR the cutoff value is updated according to the equation: $C_{i+1} = r.C_i$. We use the following equation in RDGR:

$$C_{i+1} = \begin{cases} r.C_i & \text{when the solution has improved at the } i^{th} \text{ restart} \\ C_i & \text{otherwise} \end{cases} \tag{1}$$

In RGR, the cutoff value for each restart is determined *independently* of how search performed at the previous step. However, this is not the case for RDGR. Each time search

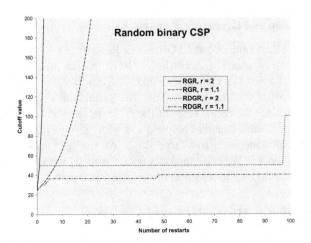

Fig. 3. Increase of the cutoff value (3 minutes).

begins with a different random seed, it traverses different search paths. Some paths may be more fruitful than others. RGR and RDGR increase the cutoff values in the same way on search paths that improve solutions. When the solution is not being improved, RGR keeps increasing the cutoff values, thus making RGR more of a randomized BT search than a randomized BT search with restarts. In contrast, RDGR maintains the cutoff value. Figure 3 shows that RGR increases the cutoff values across iterations significantly more rapidly than RDGR does, for $r=1.1$ and 2 on random binary CSPs. This explains the dynamic nature of RDGR. For problems that are not tight, solutions are found within a few restarts, and RGR and RDGR exhibit similar behaviors. For tight and over-constrained problems, RDGR seems to dominate RGR as we show in our experiments (Section 4).

4   Experiments and Results

We tested and compared the above listed 5 search strategies, namely: BT (Section 2.1), LS (Section 2.2), ERA (Section 2.3), RGR (Section 3.1), and RDGR (Section 3.2). BT is deterministic and the other 4 search techniques (i.e., LS, ERA, RGR, and RDGR) are stochastic. In the terminology introduced by Hoos and Stützle in [17], these are optimization Las Vegas algorithms, RGR is probabilistically approximately complete (PAC), and LS, ERA, and RDGR are essentially incomplete. We conducted the following three sets of experiments:

1 Effect of running time on RGR and RDGR.

Table 3. Improvements of RDGR with 95% confidence level.

Data set	Improvements over RGR		Improvements over ERA	
	Lower limit	Upper limit	Lower limit	Upper limit
1	1.16	1.61	45.16	46.77
2	1.53	1.61	-6.15	-6.15
3	3.44	3.44	27.58	31.03
4	1.85	1.85	24.07	27.77
5	0	1.85	-3.7	-3.7
6	1.56	1.56	-6.25	-6.25

of run-time in run-time distributions. The horizontal axis represents in percent the relative deviation of the solution size s from the longest known solution s_{opt}, computed as $\frac{(s_{opt}-s)100}{s_{opt}}$. Thus, the point 0% on the x-axis denotes the longest solution and, the point 20% denotes a solution that is 20% shorter that the longest solution. The vertical axis represents the percentage of test runs.

2. *Descriptive statistics* of all the solutions found, for all search techniques. This includes the measures: mean, median, mode, standard deviation, minimum, and maximum of the solution.

3. *95% confidence interval* of the mean improvement. The confidence interval was computed using the Mann-Whitney test. Table 3 reports the improvements of RDGR over RGR and ERA.

We tested these search techniques on the 6 real-world data-sets of the GTAAP of Table 1 and 4 sets of randomly generated problems. For the GTAAP data sets, we repeated our experiments 500 times for all stochastic search techniques. Naturally, a single run is sufficient for BT because it is deterministic. We found that the average run-time for all stochastic algorithms stabilizes after 300 runs on all the GTAAP data sets, as shown in Figure 4 for data set 1, which justifies our decision. We report the results for the following data sets (the same qualitative observations hold across all data sets):

– Data set 1 as a representative of an over-constrained problem.
– Data set 5 as a representative of a tight but solvable problem.

For randomly generated problems, we used the model-B-type generator of Hemert [18]. We generated three types of randomly generated problems, each containing 100 instances and each instance run for 3 minutes:

– *Under-constrained instances.* The first type of randomly generated problems are under-constrained binary CSPs with 40 variables, uniform domain size of 20 values, 0.5 proportion of constraints, and 0.2 constraint tightness.
– *Over-constrained instances.* The second type of randomly generated problems are *over-constrained* binary CSPs with 40 variables, uniform domain size of 20 values, 0.5 proportion of constraints, and 0.5 constraint tightness.
– *Instances at the phase transition.* The third type of randomly generated problems are from the *phase transition* area. These are binary CSPs with 25 variables, uniform domain size of 15 values, 0.5 proportion of constraints, and 0.36 constraint tightness. We split these instance into two sets, each of 100 instances, separating solvable instances and unsolvable instances.

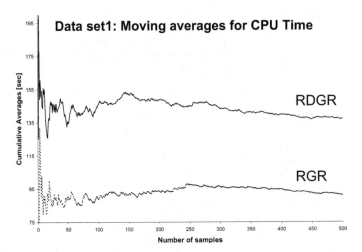

Fig. 4. Moving average for CPU run-times for data set 1.

4.1 Effect of the Running Time on RGR and RDGR

To compare the performance of RGR and RDGR, we tested them on various running times for the GTAAP data sets. The results are shown in Figures 5 and 6. In both these figures, RDGR consistently outperforms RGR over different run-times. Further, increasing the running time has no affect on the relative dominance of algorithms.

4.2 Influence of the Ratio r

We tested RGR and RDGR with different ratios, with 5 minutes running time. For the GTAAP problem we tested the values: $1, 1.1, 2^{\frac{1}{4}}, 2^{\frac{1}{2}}, 2$, and 4. For the random CSPs we tested the values: $1, 1.1, 2^{\frac{1}{4}}, 2^{\frac{1}{2}}, 2, 3$, and 4. Figures 7, 8, 9, and 10 show the influence of the ratio r used to increase the cutoff value in RGR and RDGR. In accordance with [2], Figures 7 and 9 show that a value of $r=1.1$ is the best among the values tested for RGR.

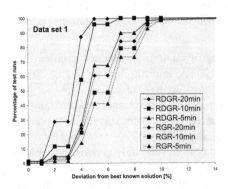

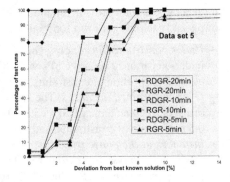

Fig. 5. Varying run time: GTAAP, over-constrained.

Fig. 6. Varying run time: GTAAP, solvable.

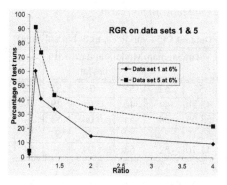

Fig. 7. Effect of r: RGR on GTAAP.

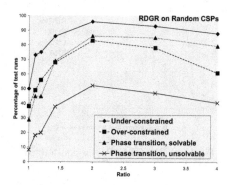

Fig. 8. Effect of r: RDGR on GTAAP.

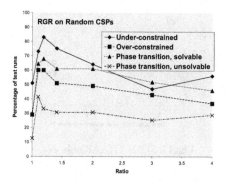

Fig. 9. Effect of r: RGR on random CSPs.

Fig. 10. Effect of r: RDGR on random CSPs.

This While, for RGR, this optimal ratio does not change with the problem type (i.e., GTAAP vs. random problem), it does for RDGR. For the GTAAP, it is $r=1.1$ (Figure 8). For randomly generated problems, it is $r=2$ (Figure 10). Our experiments indicate that the curves remains flat around these 'optima.'

4.3   Relative Performance of BT, LS, ERA, RGR, and RDGR

In this section we compare the relative performance of all the search techniques developed for the GTAAP system. Each stochastic algorithm was run 500 times of 10 min each on the GTAAP data-set, and on 100 instances of random CSPs of 3 min each. Figures 11 and 12 show the relative performance of the search techniques on GTAAP data. Figures 13, 14, 15, and 16 show the relative performance for the random problems. We do not show LS and ERA in Figure 14 because they go off the scale.

Improvement of RDGR over BT: Table 4 shows that the maximum value of the solution sizes produced by RDGR is clearly greater than that of the solution sizes produced by BT. However, due to its stochastic nature, RDGR suffers from high instability in its solution quality.

Superiority of RDGR over LS: The performance of RDGR is clearly superior to that of LS (see Table 4 and Figures 11, 12, 13, 15, and 16). Although the solution quality is

Table 4. Statistics of solution size (500 runs, 10 min each).

	Data set 1 (69 variables, over-constrained)						Data set 5 (54 variables, tight but solvable)					
Search	Mean	Median	Mode	Standard deviat.	Min.	Max.	Mean	Median	Mode	Standard deviat.	Min.	Max.
BT	57	57	57	0	57	57	52	52	52	0	52	52
LS	47.12	48	49	4.44	30	55	42.88	44	46	3.94	29	50
ERA	30.99	31	32	4.37	18	45	53.99	54	54	0.04	53	54
RDGR	59.66	60	60	0.77	58	62	52.17	52	52	0.78	50	54
RGR	58.27	58	58	2.83	23	62	51.70	52	52	1.04	49	54

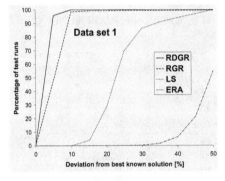

Fig. 11. SQDs: GTAAP, over-constrained.

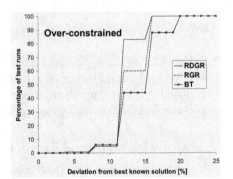

Fig. 12. SQDs: GTAAP, solvable.

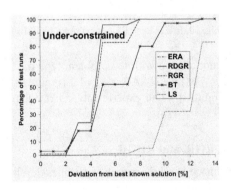

Fig. 13. SQDs: under-constrained, random CSPs.

Fig. 14. SQDs: over-constrained, random CSPs.

highly variable for both RDGR and LS, the low mean value of the solution quality of LS ensures that RDGR remains superior to LS.

Superiority of RDGR over ERA on Over-Constrained Problems: On over-constrained problems (Figure 11 and Table 3), the deadlock phenomenon prevents ERA from finding solutions of quality comparable to those found by the other techniques [10–12]. BT, LS, RDGR, and RGR do not exhibit such a dichotomy of behavior between over-constrained cases and solvable instances.

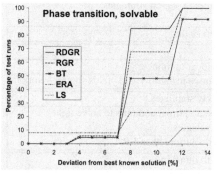
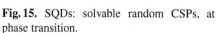
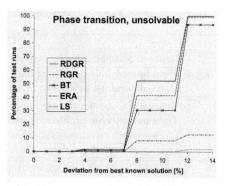

Fig. 15. SQDs: solvable random CSPs, at phase transition.

Fig. 16. SQDs: unsolvable random CSPs, at phase transition.

Performance of ERA: On solvable problem instances (Figures 12 and 13), ERA dominates all techniques. It is the only algorithm that finds complete solutions for nearly all the runs. ERA completely dominates LS. However, on over-constrained problem instances (Figure 11), RDGR, RGR, BT and LS are superior to ERA due to the deadlock phenomenon. At the phase transition (Figures 15 and 16), the behavior of ERA is independent of the solvability of the problem. ERA performs only better than LS, while RDGR, RGR and BT perform better than ERA. This difference in performance of ERA may have to do with the structure of the randomly generated problems and the GTA problem. More tests are needed to understand this phenomenon.

RDGR is More Stable than RGR: Table 5 shows the standard deviation of RGR and RDGR on the GTAAP data sets. Due to their stochastic nature, RDGR and RGR techniques show variation in their solution quality. However, the smaller standard deviations of RDGR compared to RGR in Table 5 show that RDGR is relatively more stable than RGR.

Sensitivity of LS to Local Optima: LS sensitivity to local optima makes it particularly unattractive in our context. Even BT outperforms LS.

Table 5. Standard deviation in solution quality on GTAAP data.

Data set	1	2	3	4	5	6	7	8
RGR	2.8	1.1	0.7	1.0	1.0	1.2	0.59	0.73
RDGR	0.7	0.8	0.6	0.9	0.7	1.1	0.43	0.47

Table 6. Average number of restarts on GTAAP data.

Data set	1	2	3	4	5	6	7	8
RGR	16.7	17.4	22.5	14.7	22.4	19.5	27.8	30.4
RDGR	74.5	59.9	167.4	39.1	39.1	46.2	826.2	272.0

Table 7. Comparing the behaviors of search strategies.

	Characteristics
ERA	**General:** Stochastic and incomplete
	Tight but solvable problems: Immune to local optima
	Over-constrained problems: Deadlock causes instability and yields shorter solutions
LS	**General:** Stochastic, incomplete, and quickly stabilizes
	Tight but solvable problems: Liable to local optima, and fails to solve tight CSPs even with random-walk and restart strategies
	Over-constrained problems: Finds longer solutions than ERA
RDGR	**General:** Stochastic, incomplete, immune to thrashing, produces longer solutions than BT, immune to deadlock, reliable on unknown instances, and immune to local optima, but less than ERA
RGR	**General:** Stochastic, Approximately complete, less immune to thrashing than RDGR, and yields shorter solutions than RDGR in general.
BT	**General:** Systematic, complete (theoretically, rarely in practice), liable to thrashing, yields shorter solutions than RDGR and RGR, stable behavior, and more stable solutions than stochastic methods in general

Larger Number of Restarts in RDGR: Table 6 shows the average number of restarts occurring in RGR and RDGR. This confirms our expectations stated in Section 3.2 that RDGR performs more restarts than RGR.

Summary: The following five statements, where $\succ$ denotes dominance of an algorithm over another, summarize the behavior of the 5 search strategies, also shown in Table 7:

- On unsolvable instances:
  - Beyond the phase transition: **RDGR** $\succ$ RGR $\succ$ BT $\succ$ LS $\succ$ ERA.
  - Around the phase transition: **RDGR** $\succ$ RGR $\succ$ BT $\succ$ ERA $\succ$ LS.
- On solvable instances:
  - Beyond the phase transition: ERA $\succ$ **RDGR** $\succ$ RGR $\succ$ BT $\succ$ LS.
  - Around the phase transition: two cases must be distinguished (see Figure 15). If we focus on the percentage of problems solved (i.e., lower values of SQDs), ERA remains the dominant technique: ERA $\succ$ **RDGR** $\succ$ RGR $\succ$ BT $\succ$ LS. However, if we accept larger values of the deviation from the best solution, then RDGR statistically dominates: **RDGR** $\succ$ RGR $\succ$ BT $\succ$ ERA $\succ$ LS.

5   Conclusions and Future Work

By addressing a real-world application, we are able to identify, characterize, and compare the behavior of various search techniques. BT is stable but suffers from thrashing. LS is vulnerable to local optima. ERA shows difference in performance with different problem types: while it has an amazing ability to solve under-constrained problems, its performance degrades on over-constrained problems due to the livelock phenomenon.

Restart strategies effectively prevent thrashing, but their solution quality is highly variable. RGR operates by increasing cutoff values at every restart, which increases its vulnerability to thrashing. RDGR attenuates this effect by making the cutoff value depend upon the result obtained at the previous restart, which increases the number of restarts in comparison to RGR. Consequently, RDGR exhibits a more stable behavior than RGR while yielding at least as good solutions. In the future, we plan to study the following directions:

1. Validate our findings on other real-world case-studies.
2. Design 'progress-aware' restart strategies, that is, strategies that can decide, *during* a given restart, whether to continue or abandon this particular execution.
3. Use our current application as a 'platform' to study and characterize the performance of other deterministic and stochastic search techniques.
4. Design new search hybrids where a solution from a given technique such as ERA is fed as a seed to another one such as heuristic backtrack search.

Acknowledgments

This work is supported by NSF grants #EPS-0091900 and CAREER #0133568. The experiments were conducted utilizing the Research Computing Facility of the University of Nebraska-Lincoln.

References

1. Gomes, C.P., Selman, B., Kautz, H.: Boosting Combinatorial Search Through Randomization. In: Proceedings of the Fifteenth National Conference on Artificial Intelligence (AAAI'98). (1998) 431–437
2. Walsh, T.: Search in a Small World. In: Proc. of the 16th IJCAI. (1999) 1172–1177
3. Lim, R., Guddeti, V.P., Choueiry, B.Y.: An Interactive System for Hiring and Managing Graduate Teaching Assistants. In: Conference on Prestigious Applications of Intelligent Systems (ECAI 04), Valencia, Spain (2004) 730–734
4. Glaubius, R.: A Constraint Processing Approach to Assigning Graduate Teaching Assistants to Courses. Undergraduate Honors Thesis. Department of Computer Science & Engineering, University of Nebraska-Lincoln (2001)
5. Glaubius, R., Choueiry, B.Y.: Constraint Constraint Modeling and Reformulation in the Context of Academic Task Assignment. In: Working Notes of the Workshop Modelling and Solving Problems with Constraints, ECAI 2002, Lyon, France (2002)
6. Glaubius, R., Choueiry, B.Y.: Constraint Modeling in the Context of Academic Task Assignment. In Hentenryck, P.V., ed.: 8th International Conference on Principle and Practice of Constraint Programming (CP'02). Volume 2470 of LNCS., Springer (2002) 789
7. Freuder, E.C., Wallace, R.J.: Partial Constraint Satisfaction. Artificial Intelligence **58** (1992) 21–70
8. Prosser, P.: Hybrid Algorithms for the Constraint Satisfaction Problem. Computational Intelligence **9 (3)** (1993) 268–299
9. Guddeti, V.P.: Empirical Evaluation of Heuristic and Randomized Backtrack Search. Master's thesis, Computer Science & Engineering, University of Nebraska-Lincoln (2004)

10. Zou, H., Choueiry, B.Y.: Characterizing the Behavior of a Multi-Agent Search by Using it to Solve a Tight, Real-World Resource Allocation Problem. In: Workshop on Applications of Constraint Programming, Kinsale, County Cork, Ireland (2003) 81–101

11. Zou, H.: Iterative Improvement Techniques for Solving Tight Constraint Satisfaction Problems. Master's thesis, Computer Science & Engineering, University of Nebraska-Lincoln (2003)

12. Zou, H., Choueiry, B.Y.: Multi-agent Based Search versus Local Search and Backtrack Search for Solving Tight CSPs: A Practical Case Study. In: Working Notes of the Workshop on Stochastic Search Algorithms (IJCAI 03), Acapulco, Mexico (2003) 17–24

13. Minton, S., Johnston, M.D., Philips, A.B., Laird, P.: Minimizing Conflicts: A Heuristic Repair Method for Constraint Satisfaction and Scheduling Problems. Artificial Intelligence **58** (1992) 161–205

14. Barták, R.: On-Line Guide to Constraint Programming. kti.ms.mff.cuni.cz/~bartak/constraints (1998)

15. Liu, J., Jing, H., Tang, Y.: Multi-Agent Oriented Constraint Satisfaction. Artificial Intelligence **136** (2002) 101–144

16. Luby, M., Sinclair, A., Zuckerman, D.: Optimal Speedup of Las Vegas Algorithms. In: Israel Symposium on Theory of Computing Systems. (1993) 128–133

17. Hoos, H., Stützle, T.: Stochastic Local Search Foundations and Applications. Morgan Kaufmann (2004)

18. van Hemert, J.I.: RandomCSP: generating constraint satisfaction problems randomly. homepages.cwi.nl/~jvhemert/randomcsp.html (2004)

Dynamic Distributed BackJumping

Viet Nguyen[1], Djamila Sam-Haroud[2], and Boi Faltings[2]

[1] Laboratory of Autonomous Systems,
Ecole Polytechnique Federale de Lausanne (EPFL),
CH-1015 Lausanne, Switzerland
[2] Laboratory of Artificial Intelligence,
Ecole Polytechnique Federale de Lausanne (EPFL),
CH-1015 Lausanne, Switzerland
{viet.nguyen,jamila.sam,boi.faltings}@epfl.ch

Abstract. We consider Distributed Constraint Satisfaction Problems (DisCSP) when control of variables and constraints is distributed among a set of agents. This paper presents a distributed version of the centralized BackJumping algorithm, called the *Dynamic Distributed BackJumping – DDBJ* algorithm. The advantage is twofold: *DDBJ* inherits the strength of synchronous algorithms that enables it to easily combine with a powerful dynamic ordering of variables and values, and still it maintains some level of autonomy for the agents. Experimental results show that *DDBJ* outperforms the *DiDB* and *AFC* algorithms by a factor of *one to two* orders of magnitude on hard instances of randomly generated DisCSPs.

Keywords: Search, Constraint Satisfaction, Distributed Systems, Multi-Agent Systems.

1 Introduction

Constraint Satisfaction has been used as a powerful paradigm for general problem solving. It consists of finding values for problem variables in some particular domains subject to constraints that specify possible consistent combinations. Solving a CSP is to find a set of variable assignments that satisfies all the constraints.

A distributed CSP (DisCSP) is a CSP where variables and constraints are distributed among a network of automated agents. Each agent may hold one or more variables which are connected by local constraints, and also connected by inter-constraints to variables of other agents. Many application problems in Multi-Agent Systems (MAS) can be formulated and solved using a DisCSP framework ([1]), such as distributed resource allocation problems, distributed scheduling problems or multi-agent truth maintenance tasks.

In solving DisCSPs, agents exchange messages about the variable assignments and conflicts of constraints. Several distributed search algorithms have been proposed for solving DisCSPs. They can be divided into two main groups: asynchronous and synchronous algorithms. The former are algorithms in which the process of assigning variable values and exchanging messages is performed asynchronously between the agents, whereas in the latter group, agents assign values to variables in a synchronous, sequential way. Each group has different strengths and drawbacks.

B. Faltings et al. (Eds.): CSCLP 2004, LNAI 3419, pp. 71–85, 2005.

The main contribution of this paper is to introduce the first distributed version of the centralized algorithm *BackJumping* ([2]), called *Dynamic Distributed BackJumping*. The advantage is twofold: *DDBJ* inherits the strength of synchronous algorithms that enables it to easily combine with a powerful dynamic ordering of variables and values, and still it maintains some level of autonomy for the agents. Experimental results show that *DDBJ* outperforms some existing algorithms.

2   Related Work

One of the pioneer algorithms is the *Asynchronous BackTracking – ABT* algorithm ([3,4]). It is a distributed, asynchronous version of a generic backtracking algorithm. Agents communicate by two types of messages: OK? messages to distribute the current value, and Nogood messages to declare new constraints. The simplicity and computational concurrency are its strengths. *ABT* needs polynomial space for storing nogoods to be complete ([3]). The algorithm requires the assumption that messages are received in the order in which they were sent for completeness, otherwise all nogoods have to be stored and it would suffer from exponential space complexity. One way to work around is to attach a sequence number for each message, so the order of messages can be determined at the receiving end.

A later version of *ABT* which makes use of dynamic ordering of agents, called the *Asynchronous Weak-Commitment Search – AWC*, is given in [4]. This algorithm is shown to be faster than *ABT*, but the main drawback is that it requires exponential space for completeness.

The *Distributed Dynamic Backtracking – DiDB* algorithm is another distributed, asynchronous algorithm which is inspired by its centralized version *Dynamic Backtracking* ([5]), presented in [6, 7]. The algorithm transforms the constraint network into a directed acyclic graph and performs dynamic jumps over the set of conflicting agents. This algorithm requires the assumption that messages are received in the order in which they were sent and polynomial space for nogood stores. However, the main weakness is the problem of message duplication. Due to asynchrony, an agent may keep asking values of its parents, and the parents keep sending reply messages. This process propagates down the whole graph, creates many duplicated messages. Experimental results show that the number of messages increases dramatically.

Another distributed asynchronous algorithm is given lately in [8], the *Asynchronous Aggregation Search – AAS*. This algorithm works in a similar way as *ABT*, except that consistent values of the partial solution are also included in OK messages. This mechanism helps in reducing number of backtracks. For problems with large variable domains, including consistent values produces long messages. Thus, *AAS* is more practical for problems with small variable domains.

A recently proposed algorithm, called the *Asynchronous Forward Checking – AFC* ([9]), belongs to the group of distributed synchronous algorithms. It is a generic backtracking algorithm combined with a look ahead heuristic by means of asynchronous forward checking messages. Agents assign their values for variables sequentially by having one current partial assignment shared among all agents. When a dead end is detected, the algorithm backtracks sequentially following the reverse ordering. A strength of this algorithm is in its algorithmic simplicity and good computational efficiency, inherited from centralized algorithms. It has been shown to provide better performance, in terms of number of messages and constraint checks, than asynchronous algorithms *ABT* and *DiDB* ([9]). The main drawback of *AFC* is that it does not exploit concurrency: at any time, there is only either one AFC or one BT message that is exchanged between the agents, results in long running time compared to asynchronous algorithms.

3   Preliminaries

Classically, Constraint Satisfaction Problems (CSP) have been defined for problems in centralized architectures. A finite CSP is defined by a triple $(\mathcal{X}, \mathcal{D}, \mathcal{C})$, where

- $\mathcal{X} = \{x_1, ..., x_n\}$ is the set of n variables.
- $\mathcal{D} = \{D_1, ..., D_n\}$ is the set of n finite, discrete domains of variables $x_1, ..., x_n$, respectively.
- $\mathcal{C} = \{C_1, ..., C_k\}$ is the set of k constraints on the variables. These constraints give the allowed values that the variables can simultaneously take. var(C_i) is the set of variables that are constrained by C_i.

A *solution* to a CSP is an assignment of values taken from the domains to all variables such that all the constraints are satisfied. Constraint satisfaction is NP-complete in general, and it is typically solved by a tree-search procedure with backtracking.

A distributed CSP (DisCSP) is a CSP in which the variables and constraints are distributed among a network of automated agents. Formally, a finite DisCSP is defined by a 5-tuple $(\mathcal{X}, \mathcal{D}, \mathcal{C}, \mathcal{A}, \phi)$, where $\mathcal{X}$, $\mathcal{D}$ and $\mathcal{C}$ are the same as in centralized CSP, and

- $\mathcal{A} = \{A_1, ..., A_p\}$ is the set of p agents
- $\phi : \mathcal{X} \rightarrow \mathcal{A}$ is a function that maps variables to agents

Solving a DisCSP is to find an assignment of values to variables by the collective and coordinated action of automated agents. A *solution* to a DisCSP is a compound assignment of values to all variables such that all constraints are satisfied.

In DisCSP, agents communicate with each other by sending messages. We make the following assumptions for the communication model similar to those proposed in [4]:

1. An agent can send messages to other agents iff the agent knows the addresses of the agents.
2. The delay in delivering a message is finite but random; there is no message lost.

The second assumption has been partially relaxed from the original one in [4] that also assumes that messages are received in the order in which they were sent. Some algorithms (*ABT*, *DiDB*) require this assumption to be complete. Furthermore, for simplicity and without loss of generality, we assume that:

1. ϕ is a one-to-one function; it means that each agent holds only one variable; and there are no intra-agent constraints. (In DisCSP, it is assumed that intra-agent variables/constraints can be solved efficiently by some centralized algorithm. Distributed algorithms are to focus on the *cooperative* solving techniques between *distributed* solvers (e.g. agents).)
2. $\mathcal{C}$ are binary constraints so that var(C_i) = 2, and every constraint is known by both agents involved in the constraint.

By these assumptions, the constraint network is simplified to a constraint graph where agents represent graph nodes and constraints represent graph edges.

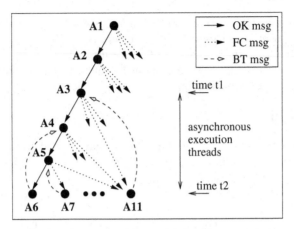

Fig. 1. An example of the *DDBJ* algorithm execution.

4   The Algorithm DDBJ

The *Dynamic Distributed BackJumping – DDBJ*, is a complete, distributed, semi-asynchronous version of a graph-based backjumping algorithm which was previously introduced in centralized CSP ([2]). The algorithm combines the concurrency of an asynchronous dynamic backjumping algorithm and the computational efficiency of the synchronous *AFC* algorithm ([9]), coupled with the heuristics of dynamic value and variable ordering.

The Distributed BackJumping Procedure

Agents perform value assignments in two phases:

- *Advancing forward* phase: which occurs when a new assignment tuple is added to the current partial solution.
- *Backjumping (backward)* phase: which occurs when an agent encounters a conflict. The process is "jumped back" to the culprit agent.

An agent is either in a *forward* phase or a *backward* phase. Algorithmically, the *forward* phase is performed sequentially: the assigning agent sends an OK to the next agent and FC messages to unassigned connected agents (similarly to *AFC* algorithm). If an agent detects a conflict when receiving some OK/FC message, it performs the *backward* phase asynchronously to backjump to the culprit agent, and also sends NG messages to unassigned agents. At any time, there can be several culprit agents detected and thus several backjumps are performed simultaneously. The culprit agents will change their values, hence the current partial solution (CPS), and perform the *forward* phase, without synchronizing with other agents nor waiting for other agents to switch phases. Consequently, at any time, agents are performing the *forward* and *backward* phases simultaneously in parallel without any synchronous control.

An example of algorithm execution is illustrated in Fig.1. At time *t1*, agent A3 sends one OK message to A4 (solid lines) and FC messages to connected agents (dotted lines). At a later time *t2*, A11 finds a conflict and backjumps to A3 by a BT message (dashed lines) and sends NG messages to others (not shown). At the same time, A3's assignment has already propagated down to A6 and A7, and get backjumped at A6 to A4 and backtracked at A7 to A5. However, the asynchronous executions at A6 and A7 and the consequent ones will soon be overwritten by the new assignment at A3. These execution flows are carried out simultaneously.

In *AFC*, backtracking is performed sequentially (or synchronously) from the detecting agent to the culprit. At any time, there is only either one OK or one BT message being sent. In *DDBJ*, any agent who receives an OK or FC message can initiate a backjump. Thus, there can be several OK and BT messages exchanged simultaneously, generating multiple asynchronous execution threads. However, there is only one OK message which may potentially lead to a solution (the most updated one or equivalently the one on the highest level of the search tree). The other OK messages will continue to propagate and create the assignment chains down the search tree, until only when the NG or newer messages arrive. Usually, it takes some cycles to stop these obsolete processes, depending on the size of the network, the connectivity density, the message delivering delay, etc.

The *DDBJ* algorithm is executed on every agent. Each maintains current value assignments of other agents in an *AgentView* ([3]). We also adopt the *AgentView-.consistent* from [6] to represent whether the CPS it holds is consistent. To determine which OK message is the most updated one and to discard obsolete messages, we introduce for each agent a time flag called *TimeStamp* which is incremented by 1 when the agent changes its value. When sending OK/FC messages, an agent includes its *TimeStamp* with its assignment. The receiving agent checks the attached *TimeStamp*s and updates its context only if the message is valid. In the example above, by the *TimeStamp*s, A4's new assignment (due to A6's backjump) will overwrite executions from A5 (due to A7's backtrack); however the new A3's assignment (due to A11's backjump) will eventually overwrite all executions below it.

The Dynamic Value and Variable Ordering Heuristics

The *DDBJ* algorithm uses dynamic value and variable ordering heuristics. Each agent keeps a potential conflict counter list of its domain values, and a potential conflict counter list of other agents (variables). An agent chooses the value which has the lowest counter value to assigns its variable, and sends the OK message (which contains the partial solution) to the agent (variable) which has the highest counter value (and FC messages to other linked agents). If there is a tie, the agent can use the chronological order. At start, all the counter values are equally zeros.

When a dead end is detected by an agent, the dead end discovering (DED) agent performs updating its priority lists in two steps. In the first step, it decreases the counter of the culprit agent (the agent whose value causes the dead end), then it sends the BT message to the culprit agent. The culprit agent, upon receiving the BT message, increases the counter of the sender (the DED agent) and increases the counter of its value that causes the backtrack, then it follows the backjumping procedure. In second step, the DED agent determines its "potential conflicting agents" (PC agents). A PC agent is the first agent whose value conflicts with a value in the domain of the DED agent. The DED agent increases the counters of the PC agents, sends a "potential conflict" – PC message to the PC agents. The PC agents, after receiving the PC message,

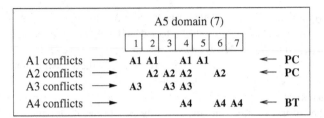

Fig. 2. An example of the heuristics: Agent A5 comes to a dead end, sends a BT message to culprit agent A4, sends "potential conflict" – PC messages to A1, A2.

increase the counters of their values (that cause the dead end), increase the counter of the DED agent. The idea here is to give more priority to the agents at higher top level of the search tree to change their values. The heuristics of dynamic ordering of value and variable would intuitively help to avoid thrashing on values selected by the very first agents and improve the ordering of agents.

An example is shown in Fig.2 to illustrate how the heuristics work. Agent A5 has 7 values in its domain. The value of agent A1 conflicts with the values (value id) 1, 2, 4, 5 of agent A5, thus these values are removed from the available values of agent A5. The value of agent A2 conflicts with the values 2, 3, 4, 6. The value of agent A3 conflicts with the values 1, 3, 4. The value of agent A4 conflicts with the values 4, 6, 7 where the value 7 is the last available value in the domain of agent A5. Thus A4 is the culprit agent with respect to agent A5. Following the first step, agent A5 increases the counter of agent A4, sends a BT to agent A4. Agent A4, upon receiving the BT, increases the counter of agent A5 and increases the counter of its corresponding value.

In the second step, agent A5 determines that A1 and A2 are the PC agents, as they are first agents who remove the values 1, 2, 3, 4, 5, 6 from its domain. Agent A3 is not a PC agent, since its value conflicts with the values 1, 3, 4 of A5 that have been removed previously by A1, A2. Thus, agent A5 increases the counters of A1 and A2, sends PC messages to A1 and A2. A1 and A2, when receive the PC message, increase the counter of A5 and also the counter of their corresponding value.

Detailed Algorithm Description

The *DDBJ* algorithm uses 8 types of messages as follows:

1. SUCCESS: a *termination* message which is broadcasted to all agents, by the last assigned agent, when a solution has been found.
2. FAILURE: a *termination* message which is broadcasted to all agents, by the first agent, when it has determined the problem has no solution.
3. ERROR: a *termination* message which is broadcasted to all agents when the algorithm encounters error (e.g. exceeded limit of time/resources).
4. OK: a message which contains the current partial solution (CPS) composed of a list of $(variable, value)$ tuples and their associate $TimeStamp$'s. This message is sent to the next agent according to the sending agent's decision of ordering.

5. FC: a message which contains a copy of OK message. This message is sent by the assigning agent to the linked agents that have not been assigned, according to its *AgentView*.
6. NG: a message which contains a nogood partial solution. It is sent to the linked agents that have not been assigned, according to its *AgentView*.
7. BT: a message which contains a nogood partial solution. It is sent back to the culprit agent (the last agent in the nogood partial solution).
8. PC: a message which contains a nogood partial solution. It is sent to potential conflicting agents determined by the agent when a conflict occurs.

The *DDBJ* algorithm is executed simultaneously on all agents in parallel. An appropriate function is called depending on the type of the received message. At start, an empty OK message is sent to the first agent for initialization.

Upon receiving an OK message, function receiveOK() is executed. It first checks if the message is valid (line 1); otherwise, it is older than, or equally timely to, the stored $TimeStamps$[1] and discarded. Next, $TimeStamps$ get updated (line 2). It then checks whether the message's partial solution (MPS) contains *the previously determined nogood* (meaning current $AgentView.consistent = false$ and the MPS contains $AgentView$). If it is the case, the agent simply does nothing and returns (line 3,4). Otherwise, it updates its context by the MPS (line 6). If the update succeeds, meaning its consistent domain of values is not empty, the agent assigns the value (line 8). Otherwise, it backtracks to the last assigned agent (line 10).

Function receiveFC() is called when an FC message is received. The agent checks and discards obsolete message (line 1), otherwise updates its $TimeStamps$ (line 2). It then checks whether the message does not contain the previously determined nogood. If it is the case, it resets the consistency state to $true$ (line 3,4). Whenever the consistency state is $true$ (line 5), the agent updates its context (line 6). If the update does not succeed, it does the following: sending NG messages to linked agents that are not assigned, sending PC messages to the determined PCAs, updating its memory of PCAs and backjumping to the culprit agent.

When receiving an NG message, the function receiveNG() checks to see if *AgentView* contains the MPS. If it is the case, it removes last one or more tuples in its *AgentView* to be the same as the received nogood, restores the values accordingly (which are associate with those tuples) (line 2) and resets the consistency state (line 3). Otherwise, if the message is newer than its *AgentView*, the agent updates its context (line 5,6,7). If the update does not succeed, it functions similarly to function receiveFC(). In both cases, if the agent is an assigned agent, it has to reset itself unassigned (line 11,12).

Function receivePC() simply updates the agent's memory of PCAs and value priority. Function receiveBT(), when a BT message is received, first updates the memory of PCAs and value priority (line 1,2). It then finds the next available value, by calling function assignVal(). Note that it has to check if the message is still valid (meaning that its variable is assigned and the message is not too old), (line 3,4,5), since several BT messages can be sent simultaneously to the agent, and some have already arrived and been processed.

---

[1] The latter happens when the agent has already received an NG message which contains the same time flag.

```
procedure receiveOK()                    .
 1: if Msg is newer than AgentView then
 2:     update TimeStamps
 3:     if previously determined nogood then
 4:         return
 5:     set AgentView.consistent = true
 6:     updateDomain(MPS)
 7:     if success then
 8:         assignVal()
 9:     else
10:         backJump(previous)
end
procedure receiveFC()
 1: if Msg is newer than AgentView then
 2:     update TimeStamps
 3:     if not previously determined nogood then
 4:         set AgentView.consistent = true
 5:     if AgentView.consistent then
 6:         updateDomain(MPS)
 7:         if not success then
 8:             update PCA
 9:             send NG to unassigned agents; PC to agents in PCA
10:             backJump(culprit)
end
procedure receiveNG()
 1: if AgentView orderly contains Msg then
 2:     restoreDom()
 3:     set AgentView.consistent = false
 4: else if Msg is newer than AgentView then
 5:     set AgentView.consistent = false
 6:     update TimeStamps
 7:     updateDomain(MPS-last)
 8:     if not success then
 9:         update PCA
10:         send NG to unassigned agents; PC to agents in PCA
11:         backJump(culprit)
12: if self is assigned then
13:     reset to unassigned
end
```

Function assignVal() tries to find a next consistent value (line 1), forwards the CPS to the next agent (line 7), otherwise it backtracks (line 9). Function backJump($Agent$-$Index$) performs the backjumping by resetting the agent context and sending BT message to agent $AgentIndex$. Function updateDomain(MPS) simply updates its value domain, $AgentView$ with the input MPS. As soon as it finds the domain empty, the function returns the detected nogood.

```
procedure receivePC()
 1: update value priority / PCA
end
procedure receiveBT()
 1: update value priority / PCA
 2: if self is assigned then
 3:     if my AgentView is NOT newer Msg then
 4:         assignVal()
end
procedure assignVal()
 1: findNextVal()
 2: if found a consistent value then
 3:     Increase TimeStamp
 4:     if self is last agent then
 5:         broadcast SUCCESS to all agents
 6:     else
 7:         send OK to next agent; FC to connected agents
 8: else
 9:     backJump(previous)
end
procedure backJump(AgentIndex)
 1: if self is first agent then
 2:     broadcast FAILURE to all agents
 3: else
 4:     set AgentView.consistent = false
 5:     reset to unassigned
 6:     send BT to agent AgentIndex
 7:     update PCA
end
```

5   Soundness, Completeness and Termination

The argument for soundness is close to the one given in [9]. The fact that agents only forward consistent assignments in OK messages at only one place in function assign-Val(), line 7, implies that the receiving agents receive only consistent assignments. A solution is reported by the last agent only in function assignVal() at line 5. At this point, all the agents have assigned their variables, and the assignments are consistent. Thus the algorithm is sound.

For completeness, we need to show that *DDBJ* is able to produce all solutions and terminate. The algorithm only backtracks, by sending BT messages, in function back-Jump(), which implements the graph-based backjumping. It has been shown in [10] that graph-based backjumping only makes *safe jumps*. In other words, the algorithm back-jumps to the culprit variable, and this jump does not lead to missing any solution. Similarly in *DDBJ*, multiple *safe jumps* may be performed at the same time simultaneously which are caused by different culprits detected by different agents. The re-assignments of the culprit agents then happen simultaneously. However, the one with the highest level in the search hierarchy tree will eventually replace all others. Thus the algorithm performs an exhaustive search and is able to produce all solutions. Hence, it is complete.

In each backtrack step, there is at least one value of a variable that is removed (line 5 in backJump()). The fact that the domains of variables are finite implies finite number of backtracks, or BT messages, until FAILURE messages are broadcasted (line 2 in backJump()). Similarly, each OK message (only sent in assignVal(), line 7) increases the number of assigned variables by 1, until the last variable where SUCCESS messages are broadcasted. Therefore, the algorithm terminates.

In *DDBJ*, agents do not have to store nogoods. An agent has to keep only the current *AgentView* and the associated *TimeStamp*'s, which have at most n elements. In addition, an agent also needs to maintain two priority lists of its value domain and other agents. Thus, the algorithm's spatial complexity is linear.

6 Experimental Results

This section gives an experimental evaluation of our algorithm *DDBJ* in comparison with two other well known algorithms, the distributed asynchronous algorithm – *DiDB* ([7]) and the distributed synchronous algorithm – *AFC* ([9]). *DDBJ* is tested in 2 versions: one version is without the dynamic ordering heuristics, called *DBJ*, to measure the performance of the semi-asynchronous backjumping procedure itself, and the other version is the full *DDBJ* algorithm.

The algorithms are tested on distributed binary CSPs which are randomly generated using the problem generator *JavaCSP* ([11]). The problems are generated based on 4 setting parameters:

- v – The number of variables (or number of agents),
- d – The number of values in the domain of each variable (domain size),
- c – The constraint density (which reflects the number of constraints), and
- t – The constraint tightness (which refers to the number of value pairs which are disallowed by the constraint).

These settings are commonly used in experimental evaluation of CSP algorithms ([12, 13, 9]). The problem generator has the ability to generate only feasible problem instances (having solutions). Thus, it is advantage to generate only feasible problem instances for problems in transition phase which are most hardest to solve and so it is easy to highlight differences in algorithm performance ([4]). Note that the problem instances are generated with the setting parameters applied globally, not by interleaving of independent subproblems.

We recall the distinction between Distributed Systems and Distributed Computing ([4]). The latter is belong to the research field of High Performance Computing, where the problem is to divide/distribute, in a efficient way, some computation load onto several connected (or distributed) computing machines. The efficiency is then defined as $speedup/N$ where N is the number of distributed machines ([14]).

In this work, we are concerning the former case, Distributed Systems, where the problems in question have their distributed characteristics in nature: they are spread over a number of distributed agents. As in [4, 7, 9, 6], we use the following measures as the criteria for evaluation:

- *Number of cycles* (or running time): to estimate the algorithm concurrency / asynchrony, as used in [3].
- *Number of messages*: to estimate the overhead of the algorithm affecting on the distributed environment, where the cost of sending messages is usually considered being more expensive than local computation of agents ([9]).
- *Number of constraint checks*: to evaluate computational efforts done locally.
- *Number of value assignments*: to represent the cost of value changes committed that may be high in some applications.

The first two measures are the most important factors in measuring the efficiency of distributed algorithms. The number of cycles indicates the running time of an algorithm. More importantly, it shows how much parallelism is exploited in asynchronous algorithms compared to synchronous ones. The notion of "concurrent checks" is discussed in [15]. In this work, we make an assumption that the constraints are simple so that an agent is able to process incoming messages, perform necessary constraint checks and send out messages in one clock cycle ([3]). Thus, the ratio "N.Constraint checks/N.Cycles" gives a good estimate of the average number of concurrent constraint checks. As argued in [16], synchronous distributed algorithms usually have better efficiency than asynchronous ones (in terms of overheads, redundant efforts, etc.), but asynchronous algorithms can exploit concurrency, thus resulting in better running time (or less number of running cycles).

The messages are set up to be delivered to destination *not necessarily in the order in which they were sent*, except for the algorithm *DiDB* where it requires the messages are delivered in order. The number of messages is an important measure for DisCSP algorithms, since in distributed environment, sending messages to other distributed agents is considered expensive ([4]).

To simulate a distributed environment and asynchronous execution, we use a discrete event simulator. We have a global discrete clock counting in cycles to simulate a real time clock. At each cycle, all agents read the incoming messages, process the computation and send out messages to other agents. If there is no incoming message, an agent simply sits idle. We recall the assumption that an agent is able to process incoming messages, perform necessary constraint checks and send out messages in one clock cycle. The algorithm is executed simultaneously in parallel on all agents. All agents terminate when an termination message is broadcasted and the algorithm finishes. The algorithm's running time is counted as the number of global clock cycles. Furthermore, to simulate the real distributed environment as close as possible, we set up the link channels between agents such that the delivery time is randomly generated between 1 and the total number of agents, which best reflects the effect of the size of the constraint network. Because the concurrency of computation of asynchronous algorithms is difficult to see from other measurements (number of constraint checks, number of messages), this setting helps to differentiate asynchronous and synchronous (or sequential) execution schema. The same argument for comparing algorithms is also pointed out in [15].

Because of limited space, the results of 2 test sets are presented. The first test set includes problems with the number of variables $n = 15$, the variable domain $d = 15$, the constraint density probability $c = 0.5$ and the constraint tightness varying from 0.1

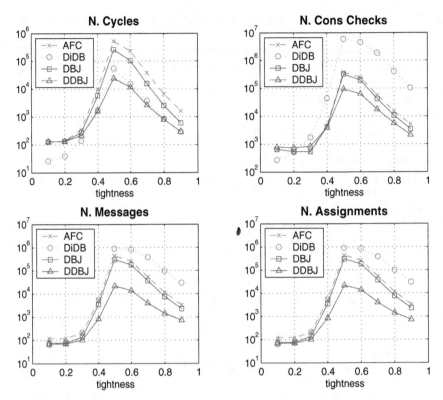

Fig. 3. Results (in *log10* scale) for N.vars v=15, domain d=15, density c=0.5. At transition phase when tightness $t = 0.5 - 0.7$, *DiDB* solved $50\% - 80\%$, *AFC*, *DBJ* and *DDBJ* solved 100% of 100 generated instances.

to 0.9 in 0.1 steps. The results in *log10* scale are shown in Figure 3. Each plot point is the average of results taken from 100 randomly generated instances. An algorithm is stopped when the number of running cycle reaches a limit of $10,000,000$ cycles or the number of messages sent *in one cycle* exceeds $100,000$.

In term of running time, *DBJ* is about 2-4 times faster than *AFC* at transition phase. The difference indicates the concurrency effect of the asynchronous backward phase of *DBJ*. *DiDB*, because of its fully asynchronous nature, is better than *DBJ* and *AFC*. However, when combined with the dynamic ordering heuristics, *DDBJ* is the best algorithm among the four for most cases.

On number of messages, *DDBJ* is better than the other three algorithms by a factor of one order approximately. The only drawback is that the message OK of *DDBJ* (and *AFC*, *DBJ*) is longer than that of *DiDB*. However, since the number of elements in a message is at most equal to the number of variables n and each element contains agent id, value id and its associate $Timestamp$, that all can be represented by 3 integer numbers, the size of a message is not more than $3n$ integer numbers.

In term of computational performance, *DDBJ* outperforms both algorithms *DiDB* and *AFC* by a factor of 5 to 100 on hard instances, where *DBJ* comes next. This can be explained by the fact that by combining good value/variable ordering heuristics and

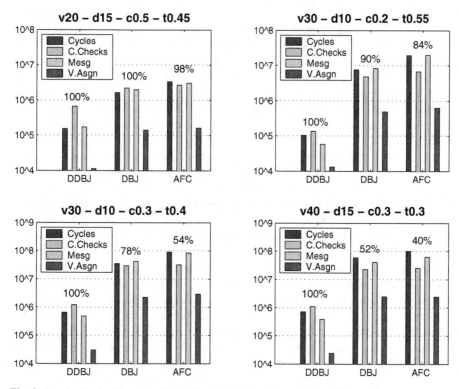

Fig. 4. Results (in *log10* scale) of feasible, high dimension problems. The percentages represent the number of problems solved within a time limit.
a) N.vars v=20, domain d=15, density c=0.5, tightness t=0.45
b) N.vars v=30, domain d=10, density c=0.2, tightness t=0.55
c) N.vars v=30, domain d=10, density c=0.3, tightness t=0.4
d) N.vars v=40, domain d=15, density c=0.2, tightness t=0.4.

exploiting concurrency, it also helps to increase the algorithm's computational efficiency and reduces the number of messages. Note that the synchronous algorithm *AFC* always performs better than the fully asynchronous algorithm *DiDB*, that it agrees with the result obtained in [9].

In more details, at transition phase where problems are hardest to solve (constraint tightness is between 0.5 and 0.7), *DiDB* is only able to solve 50% − 80% of the generated problem instances: we stop the algorithm when the number of messages sent in one cycle exceeds the limit of 100, 000 messages, since most of the time and memory resources are consumed by processing duplicated messages. This message duplication problem arises significantly when the messages are delivered with some random delay. The other three algorithms are able to solve all the problems within the limits of running cycles and messages.

In the second test set, we evaluate the algorithms by 4 feasible, high dimension problems, with the number of variables equals 20, 30, 30 and 40, respectively. The constraint tightness is set to a value close to 0.5 so that the problems are in the transition phase. The limit of number of cycles is now set to $100,000,000$. We exclude *DiDB* because of its limited capacity of solving high dimension problems: the number of messages explodes exponentially so that after a few hundred running cycles, the number of messages soon exceeds the limit of available resource. The results in *log10* scale are shown in Figure 4. The percentages show the numbers of problems solved by the algorithms. Each subgraph shows the median value of the results of 50 generated instances. The reason of taking the median value instead of the mean value is that in the transition phase, the variance of the results is too high, thus the median value indicates better the result average.

It is clear that the semi-asynchronous algorithm *DBJ* always performs better than *AFC* by a factor of 2 or more. It shows the effect of the asynchronous backjumping phase on the algorithm efficiency. *DDBJ* outperforms both the others by a factor of one to two orders for all measures. On the number of problems solved, *DDBJ* is able to solve all the problem instances for the 4 cases within the time limit, where the other two algorithms can not. This measure again confirms the high efficiency of the heuristics used in *DDBJ*. For the last two problems where the numbers of variables are 30 and 40, *AFC* is able to solve only 54% and 40% of the instances. The performance measures of *AFC* are at least one order behind those of *DDBJ*. These factors will be larger if we increase the running time limit for *AFC* to solve more instances.

One can also notice that as the number of variables increases, the performance difference between *DDBJ* and the other algorithms increases. When $v=15$, *DDBJ* is faster by about one order of magnitude, when $v=30,40$, *DDBJ* outperforms the others by about two orders of magnitude on number of running cycles and number of messages.

7   Conclusion

A new complete, distributed, semi-asynchronous algorithm, *DDBJ*, is presented. The algorithm adopts a sequentially assigning procedure, an asynchronous forward checking scheme in its *advancing phase* and an asynchronous graph-based safe-backjumping scheme in its *backjumping phase*. The sequentiality of variable assignment enables *DDBJ* to integrate the powerful heuristics of dynamic value and variable ordering and still easily to control the algorithm completeness. Experimental results show that the *DDBJ* algorithm outperforms the *DiDB* and the *AFC* algorithms by a factor of *one to two* orders of magnitude on hard instances of randomly generated DisCSPs, both on concurrent running time, number of messages and on other measures of number of constraint checks, number of variable assignments.

Acknowledgments

We would like to thank Prof. Amnon Meisels for his visiting presentation on the *AFC* algorithm. We also thank Arnold Maestre, Dr. Christian Bessière for their helpful explication of the *DiDB* algorithm. Many thanks to Dr. Bart Craenen for his problem generator *JavaCSP*. This work was performed at the Artificial Intelligence Laboratory, Ecole Polytechnique Fédérale de Lausanne and was sponsored by project COCONUT under contract number IST-2000-26063.

References

1. Yokoo, M., Hirayama, K.: Algorithms for Distributed Constraint Satisfaction: A Review. In: Proceedings of Autonomous Agents and Multi-Agent Systems. (2000)
2. Dechter, R.: Enhancement schemes for constraint processing: Backjumping, learning and cutset decomposition. Artificial Intelligence **41(3)** (1990) 273–312
3. Yokoo, M., Durfee, E., Ishida, T.: Distributed constraint satisfaction for formalizing distributed problem solving. In: Proceedings DCS. (1992)
4. Yokoo, M.: Distributed Constraint Satisfaction. Springer-Verlag (2001)
5. Ginsberg, M.: Dynamic Backtracking. Journal of Artificial Intelligence Research **1** (1993) 25–46
6. Hamadi, Y.: Interleaved backtracking in distributed constraint networks. International Journal on Artificial Intelligence Tools **11** (2002) 167–188
7. Bessière, C., Maestre, A., Meseguer, P.: Distributed Dynamic Backtracking. In: Proceedings of the IJCAI'01 workshop on Distributed Constraint Reasoning. (2001)
8. Silaghi, M., Sam-Haroud, D., Faltings, B.: Asynchronous Search with Aggregations. In: Proceedings AAAI'00. (2000)
9. Meisels, A., Zivan, R.: Asynchronous Forward-checking on DisCSPs. In: Proceedings of the Workshop on Distributed Constraints (DCR-03), Acapulco, August 2003. (2003)
10. Dechter, R., Frost, D.: Backtracking algorithms for constraint satisfaction problems – a tutorial survey. Technical report, University of California, Irvine (1998)
11. Craenen, B.: JavaCsp package. http://www.xs4all.nl/~bcraenen/JavaCsp/ (2003)
12. Prosser, P.: Binary constraint satisfaction problems: some are harder than others. In: Proceedings of the 11th European Conference on Artificial Intelligence – ECAI'94. (1994)
13. Bessiere, C.: Random Uniform CSP Generators. http://www.xs4all.nl/~bessiere/generator.html (1996)
14. Dowd, K., Severance, C.: High Performance Computing. Second edn. O'Reilly & Associates (1998)
15. Meisels, A., Kaplansky, E., Razgon, I., Zivan, R.: Comparing performance of distributed constraints processing algorithms. In: Proceedings of the Workshop on Distributed Constraint Reasoning, in AAMAS-2002. (2002)
16. Barbosa, V.C.: An Introduction to Distributed Algorithms. The MIT Press (1996)

A Value Ordering Heuristic for Local Search in Distributed Resource Allocation

Adrian Petcu and Boi Faltings

Ecole Polytechnique Fédérale de Lausanne (EPFL), CH-1015 Lausanne, Switzerland
{adrian.petcu,boi.faltings}@epfl.ch
http://liawww.epfl.ch/

Abstract. In this paper we develop a localized value-ordering heuristic for distributed resource allocation problems. We show how this value ordering heuristics can be used to achieve desirable properties (increased effectiveness, or better allocations). The specific distributed resource allocation problem that we consider is sensor allocation in sensor networks, and the algorithmic skeleton that we use to experiment this heuristic is the distributed breakout algorithm.

We compare this technique with the standard DBA and with another value-ordering heuristic [10] and see from the experimental results that it significantly outperforms both of them in terms of the number of cycles required to solve the problem (and therefore improvements in terms of communication and time requirements), especially when the problems are difficult. The resulting algorithm is also able to solve a higher percentage of the test problems.

We show that a simple variation of this technique exhibits an interesting competition behavior that could be used to achieve higher quality allocations of the resource pool. Moreover, combinations of the two methods are possible, leading to interesting results.

Finally, we note that this heuristic is domain, but not algorithm specific (meaning that it could most likely give good results in conjunction with other DisCSP algorithms as well).

Content Areas: constraint satisfaction, distributed AI, problem solving

1 Introduction

Distributed Constraint Satisfaction Problems (DisCSP from now on) are a very powerful paradigm applicable for a wide range of coordination and problem solving tasks in distributed artificial intelligence. An important subclass of these problems is the resource allocation problems, which we consider in this paper.

There is a number of distributed algorithms that were developed for this kind of problems [13] and [12] for instance. One of these, the Distributed Breakout Algorithm received quite some interest (for example [14]) because of a number of interesting properties that this algorithm exhibits (relatively simple, efficient, low overhead, linear memory requirements, good anytime characteristics).

We chose this algorithm as a basis for our work, and as a testbed we considered the sensor allocation problem described in [5, 1]. With this setup as a starting point, we then studied the effects of different search strategies on the performance of the algorithm.

B. Faltings et al. (Eds.): CSCLP 2004, LNAI 3419, pp. 86–97, 2005.

It has been shown [10] that the order in which the agents evaluate the values from their local domains plays an important role in the evolution of the algorithm towards a solution.

Our results show that by using the domain information available from their neighbors, the agents can develop search strategies that avoid resource conflicts with high probability, therefore reaching consistent assignments faster.

We see from the experimental results how one such local value-ordering technique can bring about significant improvements in terms of the number of cycles required to solve the problem (and therefore improvements in terms of communication and time requirements), especially when the problems are very difficult. The resulting algorithm is also able to solve a higher percentage of the test problems.

Moreover, a simple variation of this technique exhibits an interesting behavior that could be used to achieve higher quality allocations of the resource pool.

2   Problem Description

The distributed sensor network problem formalized in [5, 1] consists of:

- a sensor field composed of n sensors: $S = \{s_1, s_2, ..., s_n\}$
- m targets that need to be tracked: $T = \{t_1, t_2, ..., t_m\}$

Each sensor has a "range" parameter that expresses the maximum distance that it can cover; in order to successfully track a target, 3 sensors have to be assigned to that target (this is a requirement of the real sensor allocation problem: 3 sensors have to be allocated to each target in order to be able to do triangulation based on the telemetry data coming from those 3 sensors)

However, some restrictions apply:

- the sensors in the field can communicate among themselves, but not necessarily every sensor with every other sensor (the sensor connectivity graph is not fully connected). The 3 sensors tracking a given target must be able to communicate among themselves;
- any one-sensor can only track one target at a time;

2.1   Formalization

We can formalize the problem as a DisCSP assigning one agent for each target: the variables are the required sensors (three variables per agent), and the values of each variable are the sensors that can track that target (are within range).

This is a fairly general model, with multiple variables per agent and both inter and intra agent constraints, and has low inter-agent communication requirements (minimizing communication is in fact one of the goals in many real world applications). We will therefore use the terms "agent" and "target" interchangeably for the rest of the paper.

So, let's assume that we have one agent A_i for each target T_i to be tracked. This agent would then have 3 variables to control: $A_i(x_1), A_i(x_2), A_i(x_3)$; each of them is one sensor that has to be assigned to track this target. The domain of all the variables for one agent is identical (this is because sensors can be assigned to a target from the same sensor set, namely the set of sensors that can actually "see" the respective target). However, this is a very particular characteristic of the sensor network problem, and we did not make this assumption in our implementation in order to maintain generality.

In this representation of the problem, we have two types of constraints: inter-agent constraints, and intra-agent constraints.

Intra-agent constraints – the constraints within one agent:

- no two variables can be assigned the same value (one agent must have three *different* sensors tracking it)
- there must be a communication link between every two sensors that are assigned to each agent

Inter-agent constraints – the constraints between agents:

- no two variables from any two agents can be assigned the same value (one sensor can track a single target at a given time)

It is interesting to note that all constraints in this problem (except for the "visibility" ones) are constraints of mutual exclusion (typical in resource allocation problems).

3   Related Work

The idea of trying out values for the variables of a CSP in different orders, established based on various criteria, has been present in the AI literature for quite a while – e.g. [7, 8, 4, 11, 10, 6].

It has been shown that choosing the values of the variables of a CSP in an informed manner can produce significant improvements in the evolution of the search towards a solution, compared to choosing them in an arbitrary order.

Most of the existing techniques in this area are geared towards centralized mechanisms (e.g. [4]), where it is possible to achieve a global view of the current state of the problem, and establish the value-ordering based on this information. However, in a distributed setting where we perform local search, it is impossible to work under these assumptions; whatever decisions the agents may take as to the order in which they will try out their values, they must only be based on *local* information.

A further classification of these methods can be made into *static* and *dynamic* w.r.t. to when the ordering of the values is done (only in the beginning, or throughout the whole execution of the algorithm).

Dynamic ordering could in principle be expected to perform better than the static one, since it allows for more informed decisions; however, it also entails a greater runtime overhead. For example, in [10] two value-ordering heuristics are presented: a static one (NI-DBA), and a dynamic one (NPI-DBA). The authors observe that NI-DBA does not bring significant performance improvements in dense problems, however, NPI-DBA does. Therefore, we chose to compare our algorithm with NPI-DBA. It should be noted that NPI-DBA is a general-purpose heuristic (works for all types of DisCSP, not only for resource allocation).

4   Algorithms

4.1   Distributed Breakout Algorithm

The Distributed Breakout Algorithm is in fact an extension of the original Breakout Algorithm for solving CSPs in a centralized fashion [9]. This algorithm is a local search method, with an innovative technique for escaping from local minima: the constraints have weights, and the weights are dynamically increased in order to force the agents to adjust their values while in a condition of local minima.

In the distributed version, agents use *ok?* and *improve* messages for exchanging their local information: an *ok?* message is used to send the current variable value, and an *improve* message is used to send possible improvement in the evaluation of variable value. When receiving *ok?* messages from all neighbors, an agent calculates the evaluation of the current variable value and its possible maximal improvement and sends them to neighbors via *improve* messages. When receiving *improve* messages from all neighbors, an agent compares them with its own improvement. If there is a greater improvement than its own, the agent will not do anything. If there is no possible improvement (all are 0), the agent will increase the weights of the violated constraints. If its improvement is the greatest, the agent will change its variable to the value giving the maximal improvement.

Note that ties in improvement comparison are broken deterministically by comparing agent identifiers. After this step, the agents send *ok?* messages to their neighbors.

When no more constraints are violated, the problem is solved.

4.2   Preamble

We assume that the agents representing the targets all know the details of the sensor field: number of sensors, their positions and ranges.

We call two agents "neighbors" if they share a constraint. In all distributed algorithms it's necessary for each node to be able to identify its neighbors. In some cases this information is considered to be given at startup (for instance from a configuration file), and in others it is learnt at runtime (either in a "pre-processing" step, or progressively, as the algorithm runs)

In our case, we have an initial "pre-processing/discovery" phase (before we actually start DB):

- each agent determines the set of sensors that can track it (based on its coordinates, and on sensor ranges); this set will be the domain of the three local variables
- each agent sends to all his neighbors the coordinates of its target (this information is sufficient to determine the neighboring)
- upon receiving a target information from another agent, each agent determines if it has any common sensors with the respective target:
  - if so, then the agent that sent this information will be kept as a neighbor, and there will be 9 constraints of mutual exclusion between the two agents (there are 9 possible combinations of variables, and all of them have to be assigned different values)
  - if not, then the agent that sent this information will be removed from the neighbors list, and there will be no other interaction with that agent during the execution of the algorithm

- each agent sends its domain to its neighbors
- alternatively, the first step (target broadcast) could be omitted, and the second (domain broadcast) extended to all the agents: based on the domain information it is also possible to determine the neighboring

4.3 Standard Distributed Breakout Applied

Here we will present the standard DBA applied to our problem, which will be then used as a skeleton on which we build our improvements. Each agent follows Algorithm 1.1.

The differences between this version of DBA and the standard one are in the initialization phase(presented in Algorithm 1.1). There are also some changes in the send* and received* procedures made to accommodate multiple local variables, as the standard DBA allows only one variable per agent. These changes basically amount to sending and receiving the neighbors' assignments as tuples (like $X_i = < s_j, s_k, s_l >$), but they are pretty straightforward, and we don't list them here because of lack of space.

Algorithm 1.1. Standard DBA applied to sensor networks.

procedure *initialize*;
begin
    load the sensor field ;
    determine sensors "within range" → local domains;
    broadcast domain to all agents ;
    establish neighborhood based on incoming domains;
    initialize local values;
    go to standard send_values from DBA;
end
Following are the rest of the standard DBA procedures:
procedure *send_values*;
begin
    if *my improvement is best* **then** switch value ;
    if *local minima* **then** increase weights ;
    send local_values to neighbors
end
procedure *send_improvements*;
begin
    compute maximal improvement;
    send max_improve, curr_eval and curr_val to neighbors
end
procedure *received_values*;
begin
    add received values to agent_view;
    if *last message received* **then** send_improvements ;
end
procedure *received_improvements*;
begin
    record improvement;
    if *last improvement* **then** go to send_values ;
end

4.4   DBA-VO

In the standard version of the DBA, in the initialization phase, each agent randomly assigns values to its variables, and subsequently tries to assign to its variables the first values that produce a conflict reduction. The problem with this approach is that it does not take into account the fact that the initial values that the variables take can actually be very likely to cause a large number of conflicts, and that later on, a large number of cycles would be required to repair those conflicts.

The idea of DBA-VO is that if we take into account the number of times each resource (variable value, in our case) appears in the domain of the neighbors, and then try to assign each variable a value that is the least likely to cause a conflict, then it is possible that we start with an already very good assignment, that would later on require much less effort to fix. Subsequently, while trying to repair the possible conflicts, the agents would pick for their variables the values that appear least in the domains of the neighboring variables, thus reducing the likelihood that a conflict would occur in the future.

Example: let's consider the situation from Figure 1.

We see that there is a sensor S_x which is common among all the 4 targets. An uninformed assignment might look like the one in the figure, thus creating 5 conflicts (between all the agents over S_x, and another one between T_1 and T_3). Resolving these conflicts would then require 4 synchronized steps for T_1, T_2, T_3 and again T_3.

However, if the agents would have observed the fact that S_x is a highly demanded resource and therefore avoided trying to acquire it, they would not have gotten in this situation in the first place. Specifically, T_1 could have used $S_t, T_2 - S_z, T_3 - S_u$, and $T_4 - S_y$.

The information required to make these decisions is available immediately after the initialization phase, and remains valid throughout the whole execution of the algorithm.

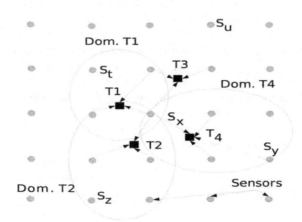

Fig. 1. Problem example.

The process is as follows:

- during the preprocessing phase, every time a new domain comes in, each agent checks to see which values from the local domain are also in the domain of the sending agent. For all such values, increase a counter.
- After the last domain comes in, sort the values from the local domain in the increasing order of the counters. Initialize the local variables with first values from the local domains.
- Afterwards, during the execution of the algorithm, every time we have a constraint violation and we have to try to find another value for the respective variable, we search for the best improving value, in the (now sorted) local domain, and pick that value, knowing that it will be likely to interfere as little as possible with the neighbors.

Intuitively, the heuristic is similar to a "non-competing contract" – agents avoid demanding a resource that they know it's likely to be busy anyway. Formally, the changes to the standard algorithm are described in Algorithm 1.2. We can easily see that all the overhead that DBA-VO has in addition to Std-DBA is basically the sorting of the domain after all the neighboring has been established (which is a one-time process, and not really expensive: $O(d \times \log d)$)

4.5 DBA-VOi

In this section we present a small variation of the previous algorithm. Namely, the heuristic works the same in as far as the domain counters are concerned; however, based on these counters, the domains of the local variables are sorted in the inverse order: the most requested ones first.

The modifications made to Algorithm 1.2 are presented in Algorithm 1.3.

The interesting effect that can be noticed after introducing this modification, is that all agents try to acquire the most "popular" resources (the ones that have the highest "demand" counters associated with them). This tendency is two-fold: first, as a result of the initialization (all agents try in the first step to acquire those resources), and second, during the subsequent conflict-repairing rounds, the agents would try to propose improvement values but again, giving preference to the most "popular" resources.

Normally, one cannot expect a gain in the running time of the algorithm when using this heuristic, exactly because of the "competition effect" described above. However, this heuristic is interesting nevertheless because it almost guarantees that a "special" subset of the available resources will be part of the final assignment. This might be important in a setting where we have for instance resources of varying quality, and we would always like to obtain as a final solution an assignment where all the best resources are in use. By simply constraining the best resources by as many consumers as possible and using this heuristic, we are then sure to obtain a final solution that respects this criteria.

4.6 Discussion

It is possible to combine the two heuristics in many ways, depending on the requirements of the domain.

Algorithm 1.2. DBA-VO.

procedure *initialize*
begin
    foreach *local variable* x_i **do**
        initialize the vector dom_cnt(x_i) with 0;

    endforeach
end
procedure *received_domain(dom)*
begin
    foreach *value* v_i *in received_domain* **do**
        foreach *local variable* x_i **do**
            if $v_i \in domain(x_i)$ **then** dom_cnt(x_i, v_i) ++ ;

    endforeach
    endforeach
    if *dom is last_domain_to_receive* **then** go to initialize_local_values ;
end
procedure *initialize_local_values*
begin
    sort values in dom(x_i) in the ascending order of dom_cnt(x_i);
    initialize local variables with the first values in their domains;
    go to standard send_values from DBA;
end
procedure *send_improvements*
in standard DBA there is a step "find improvement". we redefine this step as follows:
procedure *compute_improvements*
begin
    find local value giving best improvement; the search is done in the ascending order of
    dom_cnt(x_i);
end

Algorithm 1.3. DBA-VOi.

procedure *initialize_local_values*
begin
    sort values in dom(x_i) in the descending order of dom_cnt(x_i);
    initialize local variables with the first values in their domains;
    go to standard send_values from DBA;
end
procedure *send_improvements*
in standard DBA there is a step "find improvement". we redefine this step as follows:
procedure *compute_improvements*
begin
    find local value giving best improvement; the search is done in the descending order of
    dom_cnt(x_i);
end

If, for instance, the final assignment is important, but we would like to avoid the extra overhead generated by the continuous "fight" of the agents over the same set of "popular" resources, then we could do the initialization according to the DBA-VOi (domains sorted in the descending order of the domain counters), and continue the search in the subsequent improvement steps according to the DBA-VO heuristic (domains sorted in the ascending order of the domain counters).

Another possibility is if the time-to-solution is important, and good anytime characteristics are required. In that case, we could do the inverse: the initialization according to the DBA-VO, to start with an assignment that is as close to a solution as possible, and continue the search with DBA-VOi to go towards a solution that uses as many qualitative resources as possible.

We could even imagine a probabilistic combination of the two heuristics: for instance, while doing the initial assignments, choose for each variable a value which corresponds with high probability to the DBA-VO order, but with a small probability, choose a value corresponding to the DBA-VOi order. In this way, we would end up with a balanced initial assignment that would also have a high overall probability of making use of the qualitative resources.

5   Evaluation

A requirement of the real sensor allocation problem is that 3 sensors have to be allocated to each target in order to be able to do triangulation based on the telemetry data coming from those 3 sensors. We made our evaluations with the same settings as in [10]: the sensor field was a network of 400 sensors, and we experimented with 110 to 130 agents. This means that in total, our experiments ran with 330 to 390 variables respectively. Obviously, the problems were increasingly difficult, not only because the number of agents increased, but also because the number of required sensors (3 times the number of targets) approached the number of available sensors (total number of sensors in the grid). This made the allocation increasingly difficult, and for the 130-targets problem (which is very close to the maximum size possible), almost impossible.

This is well in line with [10], where the area around 130 targets was also shown to contain the most difficult problems of this type.

For small numbers of targets, all tested algorithms performed well; the differences start to appear only when the problems become difficult. Therefore, on the curves that we present, we show the results only from the most interesting tests, with 110 targets and more.

The problems were randomly generated, in such a way that they were solvable. However, DBA being incomplete, not all of them were actually solved. We set the maximum number of iterations that DB goes through to 50000, after which the problem was declared unsolvable.

We logged the time spent to solve each problem, the number of cycles required, and whether the problem was solved or not. We developed a visual interface that allows us to monitor the solving process.

We can see what percent of the problem instances were solved by different search strategies in Figure 2. The average number of rounds is shown in Figure 3. The average time spent for each problem size by each method is shown in Figure 4.

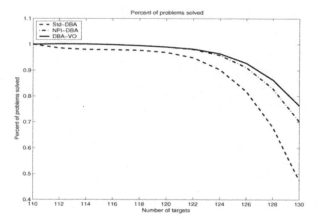

Fig. 2. Percent of problems actually solved.

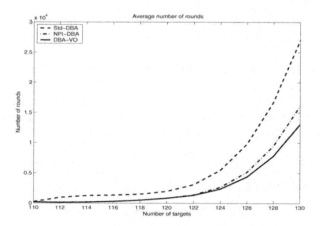

Fig. 3. Average number of rounds.

We define an empirical parameter "problem density" ρ as follows

$$\rho = \frac{number\_of\_targets \times 3}{number\_of\_sensors}.$$

This parameter will vary with the number of targets from 0 (for 0 targets) to almost 1 (for the maximum number of targets that in this case is 133).

We can clearly see in all the curves that the methods are quite similar in performance for smaller values of ρ, up to a point where ρ approaches 1. Figure 2 shows that there is a steep decrease in the percentage of the problems solved by all algorithms, but DBA-VO performs best in that area (manages to solve most of the problems), followed by NPI-DBA (about 70%), and standard DBA (less than half of the problems solved).

In figure 3 we see that on average, DBA-VO does less than half of the rounds of Std-DBA, and about 25% less rounds than NPI-DBA.

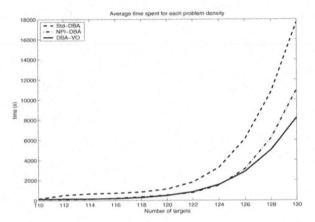

Fig. 4. Time spent for each problem size.

Based on these results, we can conclude that both the "informed" initialization of the variables and the subsequent search strategy plays a role in the performance of the algorithm.

We also recorded the time required to solve each problem by the different methods, having in mind the fact that as little as may be, there is an overhead in DBA-VO that standard DBA does not have. However, similar to the number of rounds, we can see in figure 4 that this overhead pays off eventually, and we achieve better results.

Overall, we see that DBA-VO outperforms its counterparts in all the three considered measures.

6   Conclusions and Future Work

We presented a value-ordering heuristic for improving the performance of the Distributed Breakout Algorithm applied on distributed resource allocation problems.

We compared this technique with the standard DBA and with another value-ordering heuristic [10] and saw from the experimental results that it outperforms both of them in terms of the number of cycles required to solve the problem (and therefore improvements in terms of communication and time requirements), especially for difficult problems. The resulting algorithm is also able to solve a higher percentage of the test problems.

Moreover, a simple variation of this technique exhibits an interesting behavior that could be used to achieve higher quality allocations of the resource pool (ensuring that a certain subset of the resources is allocated in a final assignment). Interesting combinations of the two techniques are possible, giving desirable properties of the allocation algorithm.

Further improvements could be obtained by allowing multiple simultaneous changes of the local variables at each step, or by trying a hierarchical approach to the problem, where certain agents are delegated as a "local authority" for solving a particularly difficult local problem.

It would be interesting to study in more detail the performance gains brought by combinations of these techniques when the problem size increases, in terms of two dimensions: the size of the sensor field (thus also the maximum number of targets), and the sensor ranges (thus the size of the domains).

As future work we also plan to compare our algorithms with other distributed methods for constraint satisfaction as the focal point techniques from [2], or the distributed stochastic search from [3].

References

1. R. Bejar, B. Krishnamachari, C. Gomes, and B. Selman. Distributed constraint satisfaction in a wireless sensor tracking system, 2001.
2. M. Fenster, S. Kraus, and J. Rosenschein. Coordination without communication: Experimental validation of focal point techniques. In *ICMAS 95*, 1995.
3. S. Fitzpatrick and L. Meertens. An experimental assesment of a stochastic, anytime, decentralized, soft colourer for sparse graphs. In *1st symposium on stochastic algorithms: foundations and applications*, pages 49–64, 2001.
4. D. Frost and R. Dechter. Look-ahead value ordering for constraint satisfaction problems. In *Proceedings of the International Joint Conference on Artificial Intelligence, IJCAI'95*, pages 572–578, Montreal, Canada, 1995.
5. C. Gomes, C. Fernandez, R. Bejar, and B. Krishnamachari. Communication and computation in discsp algorithms. In *Proceedings of the Ninth International Conference on Principles and Practice of Constraint Programming (CP'02)*, pages 40–45, Ithaca, NY, USA, 2002.
6. K. Kask, R. Dechter, and V. Gogate. New look-ahead schemes for constraint satisfaction. In *In The Eighth International Symposium on Artificial Intelligence and Mathematics*, pages 998–1003, Detroit, MI, 2004.
7. N. Keng and D. Yun. A planning/scheduling methodology for the constrained resource problem. In *Proceedings of the 11th International Joint Conference on Artificial Intelligence, IJCAI-89*, pages 998–1003, Detroit, MI, 1989.
8. S. Minton, M. D. Johnston, A. B. Philips, and P. Laird. Minimizing conflicts: A heuristic repair method for constraint satisfaction and scheduling problems. *Artificial Intelligence*, 58(1-3):161–205, 1992.
9. P. Morris. The breakout method for escaping from local minima. In *Proceedings of the National Conference on Artificial Intelligence, AAAI-93*, pages 40–45, Washington, DC, 1993. AAAI Press.
10. A. Petcu and B. Faltings. Applying interchangeability techniques to the distributed breakout algorithm. In *Proceedings of the 19th International Joint Conference on Artificial Intelligence, IJCAI-03*, Acapulco, Mexico, 2003.
11. N. M. Sadeh and M. S. Fox. Variable and value ordering heuristics for the job shop scheduling constraint satisfaction problem. *Artificial Intelligence*, 86(1):1–41, 1996.
12. M. Yokoo and K. Hirayama. Distributed breakout algorithm for solving distributed constraint satisfaction problems. In V. Lesser, editor, *Proceedings of the First International Conference on Multi–Agent Systems*. MIT Press, 1995.
13. M. Yokoo and K. Hirayama. Algorithms for distributed constraint satisfaction: A review. *Autonomous Agents and Multi-Agent Systems*, 3(2):185–207, 2000.
14. W. Zhang and L. Wittenburg. Distributed breakout revisited. In *Proceedings of the National Conference on Artificial Intelligence, AAAI-2002*, pages 352–357, Edmonton, Alberta, Canada, 2002.

Automatically Exploiting Symmetries
in Constraint Programming

Arathi Ramani and Igor L. Markov

Department of EECS, University of Michigan,
1301 Beal Avenue, Ann Arbor, MI 48109, USA
{ramania,imarkov}@eecs.umich.edu

Abstract. We introduce a framework for studying and solving a class of CSP formulations. The framework allows constraints to be expressed as linear and non-linear equations, then compiles them into SAT instances via Boolean logic circuits. While in general reduction to SAT may lead to the loss of structure, we specifically detect several types of structure in high-level input and use them in compilation. Linearity is preserved by the use of pseudo-Boolean (PB) constraints in conjunction with a 0-1 ILP solver that extends common SAT-solving techniques. Symmetries are detected in high-level constraints by solving the graph automorphism problem on parse trees. Symmetry-breaking predicates are added during compilation. Our system generalizes earlier work on symmetries in SAT and 0-1 ILP problems. Empirical evaluation is performed on instances of the social golfers and Hamming code generation problems. We show substantial speedups with symmetry-breaking, especially on unsatisfiable instances. In general, our runtimes with the specialized 0-1 ILP solver Pueblo are competitive with results recently reported for ILOG Solver.

1 Introduction

Traditional constraint programming (CP) techniques such as generalized arc consistency (GAC) are frequently the methods of choice for hard problems arising in real-world applications. Well-known packages such as ECL^{i}PS^e [22] and ILOG Solver [27] offer powerful environments for constraint specification and solver deployment. These systems provide for the development of problem-specific solvers using the best available techniques for a given problem. Another option is *reduction* – a problem for which no solver is available can be reduced to one for which a solver does exist.

Boolean satisfiability (SAT) is commonly used in problem reductions, since it is well-known and many SAT solvers are available in the public domain. Unfortunately, in most cases reduction-based methods are not competitive with CP approaches developed for a problem. While CP-based techniques can take advantage of problem-specific bounds to retain tighter control of the search, SAT solvers cannot. This disadvantage is mitigated to some extent by recent breakthroughs in SAT-solving. With new exact SAT solvers such as ZChaff [19], the size and scope of application-derived instances that can be solved has widened [20]. However, many applications do not benefit from breakthroughs in SAT solving due to inefficiencies introduced while producing SAT encodings. The CNF format used for SAT instances is very restrictive, and even encoding

B. Faltings et al. (Eds.): CSCLP 2004, LNAI 3419, pp. 98–112, 2005.

simple linear constraints can result in a blowup in size. Another cause of inefficiency is the *loss of structure during problem reductions*. Examples of structure in constraints include *linearity* and *symmetry*.

The presence of symmetries slows down search due to the existence of redundant search paths. The work in [9] describes how symmetries are detected in a SAT instance by reduction to graph automorphism and broken by adding lexicographic ordering constraints, called MinLex symmetry-breaking predicates (SBPs). The addition of these SBPs accelerates SAT solvers. In [14], symmetry-breaking ordering constraints are proposed for CSPs with *matrix models*. Linear "counting" constraints popular in applications are studied in [2]. These constraints can be efficiently expressed using ILP, where linear equations are allowed, but expressing them in CNF may be expensive. On the other hand, generic ILP solvers such as CPLEX are sometimes not competitive with leading-edge SAT solvers for Boolean constraints. Linearity can be preserved using 0-1 ILP, a problem closely related to SAT but with an ILP-like input format. Specialized techniques developed for SAT can be adapted to 0-1 ILP without paying any penalty for generality. Recently, several specialized 0-1 ILP solvers such as PBS [2], Galena [7] and Pueblo [25] have been introduced. Symmetry-breaking techniques from [9, 1] were extended to 0-1 ILP in [4].

This work contributes a framework for structure-aware compilation of a class of constraint programming problems by reduction to SAT and 0-1 ILP. We generalize techniques proposed in [9, 4] to detect symmetries in high-level constraints via reduction to *graph automorphism*. Our system facilitates comparison of different encoding strategies and SAT reductions. This is useful since recent work [28, 5, 6] has shown that the encoding used can dramatically affect search speed. Our goals here are (1) to generalize earlier work on the detection of structure in SAT instances so that it is applicable to a larger class of high-level CSPs (2) to automate the task of structure-aware reduction to SAT/0-1 ILP (3) to use this framework to study the performance of structure-aware reduction techniques. Unlike earlier work [9, 2], our framework detects structure in high-level input *before reduction* and uses it to produce more effective encodings. Our empirical results for the social golfer and Hamming code generation problems show that breaking symmetries during reduction considerably improves the performance of both SAT and 0-1 ILP solvers. On many instances, our runtimes are competitive with results reported using ILOG Solver [27] in [14]. Symmetries detected by our method can be used by *any* constraints solver, not just one that assumes reduction to SAT, since we detect symmetries in high-level input. While we add SBPs during preprocessing, there are several methods that focus on breaking declared symmetries during search [24, 12] that can make use of the symmetries we detect.

The rest of the paper is organized as follows. Section 2 discusses background and previous work. Section 3 explains how symmetries are detected and broken in high-level constraints. Section 4 discusses more comprehensive symmetry-breaking, with empirical results in Section 5. Section 6 concludes the paper. The details of compilation to SAT and 0-1 ILP and the encodings we use are discussed in the Appendix.

2   Background and Previous Work

Boolean Satisfiability (SAT). A SAT instance consists of a set of 0-1 variables V, and a set of clauses C, where each clause is a *disjunction of literals*. A literal is a variable or its complement. The SAT problem asks to find an assignment to the variables in V that satisfies all clauses in C, or prove that no such assignment exists.

0-1 ILP. 0-1 ILP allows a CNF formula to be augmented with Pseudo-Boolean (PB) constraints, or linear inequalities with integer coefficients of the form: $(a_1x_1 + a_2x_2 + \ldots + a_nx_n \leq b)$ where $a_i, b \in Z$; $a_i, b \neq 0$; x_i are Boolean literals.

CNF vs. 0-1 ILP. Recent work has shown that formulating problem instances as 0-1 ILP instead of SAT can result in faster search. Specialized 0-1 ILP solvers have been developed in [2, 7, 25], and have been shown to perform better than both leading-edge SAT solvers [19] and generic ILP solvers such as CPLEX on some 0-1 ILP formulas. However, this is not always the case. For an application, there can be several reductions to SAT, and some encodings are more difficult to solve than others. CNF encodings for circuit layout applications in [2] contain large numbers of symmetries, increasing their difficulty. In [28], Warners proposes an efficient encoding where a PB constraint is replaced by a linear number of CNF clauses. In [5], a tree-based linear conversion is proposed to translate 0-1 ILP constraints to CNF. More recently, [6] discusses a GAC-preserving encoding, with a solver modification that results in SAT instances that are solved faster than their 0-1 ILP counterparts. Our approach constructs a parse tree and instantiates Boolean circuits for addition, multiplication and subtraction. Most previous work performs reduction to SAT on a per-problem basis, but we provide a high-level specification language in which constraints can be easily expressed and conversion to SAT/0-1 ILP is automated for all problems. Given the impact that efficient encodings have on search speed, our framework is designed so that different encodings can be easily plugged in and used with our symmetry-breaking infrastructure.

Symmetry Detection and Breaking. A *symmetry* of a discrete object is a reversible transformation of its components that leaves the object unchanged, e.g., permutations of graph vertices that map edges into edges. Symmetries occurring in a SAT instance indicate the presence of redundant search paths, and breaking symmetries can reduce search time. Detection of symmetries in CNF formulas by reduction to graph automorphism is proposed in [9]. A graph is built from a CNF formula such that there is a one-one correspondence between symmetries of the formula and the graph. The graph automorphism software Nauty [16] is used to detect graph symmetries. The symmetry group induces an equivalence relation on the set of variable assignments for a CNF formula. *Lex-leader* symmetry-breaking predicates (MinLex SBPs) that allow only the *lexicographically smallest* assignment in an equivalence class are defined in [9] . A more efficient SBP construction is proposed in [3]. Symmetry detection via graph automorphism is extended to 0-1 ILP in [4]. Our work generalizes these methods to a broader class of problems that use integer coefficients, non-binary variables and non-linear operations. Symmetries are detected at a higher level, eliminating the risk that some symmetries may be obscured during reduction. In [14], the author defines high-level lexicographic (MinLex), anti-lexicographic (anti-Lex) and multiset ordering constraints for CSPs with matrix models that exhibit symmetry. However, row and column symmetries must first be identified in matrix models for individual problems and constraints designed accordingly. Our system allows symmetries to be automatically detected in any problem instance, not just a matrix model, and used by any solver. This functionality may be useful to methods that focus on declared symmetries during search. A modified search procedure that performs partial symmetry-breaking for matrix models is proposed in [24], where SBPs are specified for a stabilizer set that is a subgroup of the symmetry group. We find generators of the symmetry group using the graph automorphism program Saucy [10], and these generators can be used by the algorithms in [24] to compute SBPs. Another related work is [12], which takes as input some generators of the symmetry group and uses them to check for dominating elements in the search tree. Since our system automatically detects generators it may be applicable to such algorithms. At present, we use only MinLex SBPs from [9]. We have not yet studied other types of SBPs such as those in [14]. Symmetries in linear programming problems have also been discussed in [17].

3   Symmetry Detection

Earlier work [9, 4] detects symmetries in SAT/0-1 ILP instances *after* reduction. Our approach is to detect symmetries in the high-level specification of constraints, where they correspond directly to symmetries of the formula and can be used by multiple solvers. Symmetries detected in a SAT instance can only be used by SAT solvers, or must be traced back to the original instance to understand their significance. Also, some symmetries may be obscured during reduction. For example, counting constraints are symmetric, but the most compact encodings for these constraints [28] use comparator circuits which are not symmetric.

Detecting symmetries in CNF and 0-1 ILP via graph automorphism was first proposed in [9]. We follow a similar approach for high-level symmetry detection. A parse graph is built from the constraints such that there is a one-to-one correspondence between the symmetries of the constraints and the graph symmetries. We describe the graph construction only for the arithmetic operators '+', '-', and '*', but it can be extended to include more arithmetic or logical operators by adding more colors. An example formula in our specification language and the corresponding graph construction are shown in Figure 1. The formula declares two 3-bit integers x_1 and x_2, and the constraint $x_1{}^2 + x_2{}^2 == 25$. The specification language we use is described in the Appendix. Vertex shapes in the figure indicate different colors. The figure shows the symmetry between vertices for x_1 and x_2.

The graph construction is outlined as follows.

Step 1. Each binary variable x_i in a formula is represented by two positive and negative literal vertices, v_i and v_i', which are given the same color. v_i and v_i' are connected by an edge to ensure Boolean consistency. Each multi-bit variable x_j is represented by a single variable vertex v_j. A unique color is associated with each bit size.

Step 2. For each constraint C_i, two vertices T_i and R_i represent the constraint type ($\leq, \geq, ==, !=$) and RHS value respectively. A unique color is associated with each constraint type and RHS value. The vertices T_i and R_i for a constraint C_i are connected by an edge. Additionally, for each C_i:

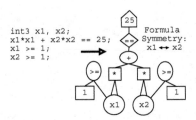

```
int3 x1, x2;
x1*x1 + x2*x2 == 25;
x1 >= 1;
x2 >= 1;
```

Formula
Symmetry:
x1 ↔ x2

Fig. 1. Constraints declaration in our specification language and the corresponding parse graph. Vertices are shaped differently to indicate different colors.

Step 2a. Variables/literals are grouped by the priority of operations in which they occur. Multiplication between variables or by coefficients has the highest priority. '+', '-' and '*' operators have distinct colors. Each distinct coefficient value in the formula is also given a unique color. Variables connected by a '*' operator are grouped under a single *coefficient vertex* that represents the product of their coefficients (if the product is unity, this vertex is omitted). This coefficient vertex is in turn attached to a *multiplication vertex*. Variables/literals not involved in multiplication operations are grouped by coefficient, with all variables having the same coefficient value connected to a common coefficient vertex.

Step 2b. After grouping multiplicative terms, we have single variables/literals or multiplicative groups connected by '+' or '-' operations. Variables/groups associated with a '+' sign are connected directly to the constraint type vertex T_i ('+' is the default operation, so there are no special vertices for it). Variables/groups associated with a '-' operation are connected to a *negation* vertex to indicate subtraction. The negation vertex is connected to the type vertex T_i.

Theorem 3.1. *Assume that a colored parse graph is constructed from a given formula of constraints as outlined above. Then, the symmetries of the constraints correspond one-to-one to the symmetries of the graph.*

Proof. We first prove that *a symmetry in the constraints is a symmetry in the parse graph.* Consider a formula with a set V of formula variables and a set C of constraints. Consider two variables, $v_1, v_2 \in V$, and let $C_1, C_2 \subset C$ be the sets of constraints that v_1 and v_2 occur in respectively. Let v_1 and v_2 be symmetric. Then, for every constraint c in C_1 there is a corresponding constraint in C_2 that is its symmetric image.

We construct a colored parse graph $G(X, E)$ for the formula where X is the set of vertices in the graph and E the set of edges. Let x_1 and x_2 be the vertices created for v_1 and v_2 respectively, and E_1 and E_2 be the edges incident on x_1 and x_2. Assume that x_1 and x_2 are *not* symmetric in the graph construction. For this to be true, it must be true that the edge sets E_1 and E_2 are not symmetric. Without loss of generality, assume there exists some edge $e \in E_1$ that does not have an image in E_2. From the graph construction rules, an edge can connect a variable vertex to one of the following: (i) a complementary literal (ii) a constraint type vertex (for addition with unit coefficient) (iii) a negation vertex (for subtraction with unit coefficient) (iv) a multiplication vertex (for multiplication with unit coefficients) and (v) a coefficient vertex that is connected to a multiplication/negation/constraint type vertex. In the first case, assume that e connects x_1 to a complementary literal vertex, and x_2 does not possess such an edge. Then, v_2 is not a binary variable, and it cannot be symmetric to v_1. In the second case, e indicates the presence of a constraint $c \in C_1$ where v_1 is added with a coefficient of 1. Since v_1 and v_2 are symmetric in the formula, there *must* be a constraint in C_2 that matches c. However, if such a constraint existed, there would be an edge representing it in E_2, symmetric to e. The same argument applies to cases (iii) and (iv). The only special case occurs in (v), when variables are multiplied together with different coefficients. We use the product of all coefficient values as the resulting coefficient. This reflects the fact that multiplication is commutative, i.e. $(av_1)(bv_2) = (ab)(v_1)(v_2)$ and $(cv_3)(dv_2) = (cd)(v_1)(v_2)$, so if $ab = cd$ then the expressions are symmetric.

For the other direction, we note that symmetries in the parse graph can only exist between vertices of the *same color*. Additional vertices are created to represent operations, but they can never be mapped to variable vertices. Thus, the only spurious symmetries we need to consider are between *variable vertices of the same bit size*. It is clear that the proof for the forward direction can be reversed for this case, i.e. edge sets incident on both vertices must be symmetric and represent symmetric constraints in the formula.

□

Avoiding Abstraction Overhead. Our graph construction generalizes earlier work in [9, 4] for CNF and 0-1 ILP formulas. Often, generalization involves paying a performance penalty – in this case, dealing with a more expressive input format that includes non-linear constraints can introduce additional vertices. This penalty can be avoided by modifying the graph when special cases are detected. Consider the case where an instance contains *only* 0-1 ILP constraints with no non-linear operations and only 1-bit variables. IN this case, our construction is designed to mimic the construction in [4], and produce *exactly* the same graphs. For pure CNF formulas, some modification is required to produce graphs as compact as the specialized constructions from [9, 1]. Since there are no coefficients or RHS values, constructions in [9] and [1] use only two types (colors) of vertices: literal and clausal. A clause with > 2 literals is represented by a clausal vertex, connected to its literal vertices. Binary clauses are represented by an edge between both literals. Graphs created by our system require constraint type and RHS value vertices for each constraint. However, CNF formulas are easy to detect. A CNF formula involves *only* binary variables. All coefficients are unity. Clauses can be expressed in two ways: as the logical-or ("$\|$") of literals, or as the additive constraint that the sum of literals must be ≥ 1. These characteristics can be tested for, and graph construction altered accordingly.

Symmetry-Breaking Predicates (SBPs). The parse graph is analyzed for symmetries using the efficient automorphism program Saucy [10], which returns generators of the symmetry group. We generate high-level lex-leader SBPs from the generators, and add them as constraints to the original instance. These SBPs are also compiled into SAT. For multi-bit variables, SBPs may be large and complex if a generator has several cycles (for a detailed description of cycles in a generator, and the resulting predicates, see [9]). We break only the first few (1 or 2) cycles in multiple-cycle generators for simplicity. For binary variables, we implement the efficient linear-sized SBP construction in [3] and add these SBPs to the CNF formula. The problems we test here all use matrix models with binary variables. The design of efficient SBPs for multi-bit variables is a direction for future research.

4   More Comprehensive Symmetry Breaking

This section discusses extensions to increase the system's coverage of symmetries.

Symmetries in Associative Expressions. Many of the operators that we support, such as '+' and '*' are associative, i.e. $x_1 + x_2 + x_3 = x_2 + x_3 + x_1$ and $(x_1 + x_2) + x_3 = x_1 + (x_2 + x_3)$. However, parse trees built from constraints often do not reflect this symmetry. In parsing, language rules are recursively matched. This imposes a non-symmetric structure on the parse tree. We avoid this non-symmetric structure by grouping all variables connected by an associative operation together. Symmetry in associative operations can also be missed when nested parentheses are used. Our system currently does not support the nesting of expressions through the '(' and ')' operators, but can be easily extended to do so. Detecting symmetries in associative operations has been addressed in the CGRASS system [11]. However, CGRASS detects symmetries in an ad-hoc way, by keeping track of the number and type of constraints a variable occurs in and matching these for different variables. Detection via graph automorphism is more comprehensive, and given efficient software such as Saucy, incurs hardly any overhead. Our method, like CGRASS, is not complete – it uses only the generators of the symmetry group found by Saucy. For complete symmetry-breaking, the full group would have to be reconstructed from the generators. This has been found to be very time-consuming [9], whereas using only generators is more efficient and often just as effective. CGRASS also undertakes simplification of constraints in other ways, which our system does not cover.

Consider the expressions $x_1 + (x_2 + x_3) + x_4$ and $x_1 + (x_2 + (x_3 + x_4))$, which are the same, but are evaluated differently due to parentheses. The order of evaluation imposed by parentheses hides the symmetry between variables, since expressions enclosed within '()' symbols are treated as separate sub-expressions. However, it is possible to simplify high-level input so that such symmetry is preserved. We list simplification rules for the operators '+', '-' and '*'.

Rule 1. Nested () symbols must be simplified before the outermost () operation. can be simplified.

Rule 2. If an expression within () symbols is flanked by '+' and '-' operations on the left and right sides, parentheses are unnecessary, e.g., in $\ldots + (x_1 + x_2) + \ldots$ the () operators can be ignored.

Rule 3. If an expression within () symbols is multiplied by a single term, the resulting expression can be evaluated, e.g., $x_2 * (x_1 + x_4)$ is written as $x_2 * x_1 + x_2 * x_4$. It is possible to simplify the parenthesized products, e.g. $(x_1 + x_2) * (x_3 + x_4)$ by implementing multiplication rules, but this may cause a size blowup in graphs for large expressions.

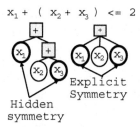

Fig. 2. Associative symmetry with parenthesized sub-expressions: x_1 and x_3 are symmetric but the original parse tree is asymmetric.

The above list of rules can be extended further, but it already facilitates the detection of symmetries in simple associative expressions. This is illustrated in Figure 2, where x_1 and x_3 are symmetric, but the symmetry is not visible in the parse graph. With the proposed modifications the associative symmetry is preserved. Our system already implements this feature for '+' and '-' operations without parentheses, where we ignore the order in which the operations occur.

Value Symmetry. We detect *formula symmetries*, that are determined by the occurrence of variables in constraints. However, *value symmetries* that occur between the actual domains of variables can also be significant. Ordering constraints for declared value symmetries are discussed in [14], and [15] describes an algorithm to detect and break value symmetries during search. We discuss how our system may be extended to detect value symmetry.

Value symmetry can arise from *operators* that control the value of a variable, e.g. the *complement* operation on binary variables: $a' = 1 - a$. The mapping $a \leftrightarrow a'$ is known as a *phase shift symmetry*. In [1], the construction from [9] is modified to detect phase-shift symmetries in almost all cases. For the non-binary case, such symmetries may arise in problems with a *cyclic* nature, e.g., scheduling problems. Any scheduling solution for {Monday, Tuesday, Wednesday} can often be shifted to {Tuesday, Wednesday, Thursday}. Such shifts can also be described by an operator – if a variable's domain is a cyclic group modulo 4, we can say $a'' = (a + 1)\%4$. Intuitively, the graph construction to represent a cyclic group of values is a cycle of vertices. However, if the domain size is > 2, this will result in spurious symmetries if all vertices are given the same color. Each vertex in the cycle must be given a different color for this construction to work. However, this prevents the detection of symmetries between values in the domain of the same variable. A set of constraints satisfied when $a = 0$ may also be satisfied when $a = 2$. This type of symmetry-detection is addressed in [15]. Adapting our techniques to detect such symmetries is more difficult, since it may require the enumeration of variable and constraint values in the graph, resulting in very large and complex graphs. Another focus of our current work is developing efficient graph constructions for this case.

5   Empirical Results

We test our system on constraint programming problems with *matrix models* with row and/or column symmetries from [14]. Each problem is modeled using the constraints described in [14] and specified in our system's input language, followed by symmetry detection and compilation to SAT and 0-1 ILP. SBPs are added to the CNF or ILP instances. We use Saucy [10] to detect symmetries, ZChaff to solve SAT instances, and the new 0-1 ILP solver Pueblo [25] to solve 0-1 ILP instances. We show results for the balanced incomplete block design problem (BIBD), social golfer problem (SG) and Hamming code generation (HC) problems. Results here are obtained using a Intel Pentium processor processor at 1GHz for the SG and HC problems, and an Intel Xeon dual processor at 2 GHz. Both systems have 1GB of RAM and run RedHat Linux 9.0. ZChaff and Pueblo runtimes are the average of 3 starts. Timeout is set at 600 seconds. For BIBD instances, we use the Xeon processor at 2GHz to compare our encodings with those in [23]. For SG and HC instances, we use the 1GHz Pentium processor to allow runtime comparisons with [14]. Symmetry-breaking ordering constraints in [14] are implemented using ILOG Solver and tested on a 1 GHz Pentium processor running Windows XP. We note that [14] also reports a "number of failures" metric, which is the number of incorrect decisions made by Solver at nodes in the search tree. We do not have access to Solver and the SAT/0-1 ILP solvers we use do not report such a statistic. However, we use exactly the same hardware as [14] so that runtime comparisons are fair. Since it is not possible for us to use Solver, we use results directly from [14].

Balanced Incomplete Block Design Problem (BIBD). This problem asks to find $b > 0$ subsets of a set V of $v \geq 2$ elements such that each subset contains exactly k elements $(v > k > 0)$, each element appears in exactly $r > 0$ subsets, and each pair of elements appears together in exactly $\lambda > 0$ subsets. An instance is expressed as (v,b,r,k,λ), and named bibd(v,b,r,k,λ) in the results table. We use the matrix model described in [14] (originally from [18]). We initially tested encodings with and without SBPs using ZChaff and Pueblo on the large instances used in [14] (originally from [8]). However, our observation on these instances was that adding MinLex SBPs actually affects performance negatively for the Pueblo solver (ZChaff is unable to solve most instances within the time limit, with or without SBPs). For satisfiable instances, this is not unusual and has been noted earlier in [9]. When there are several solutions, adding SBPs may prevent some solutions from being found earlier in the search. However, this does not explain the poor performance on unsatisfiable instances of this problem, which may be because MinLex SBPs are not useful in this case. In [14], several types of SBPs are tested, with anti-Lex constraints being most effective for BIBD. The anti-Lex SBPs are the reverse of MinLex orderings, and permit different assignments than MinLex. We can, however, use this problem to illustrate the importance of efficient encodings. SAT encodings for the BIBD problem have been developed in [23], where the instances used

Table 1. ZChaff results and Saucy statistics for BIBD instances using our encodings and those in [23], with and without SBPs. T/O indicates timeout at 600s. Pueblo is not tested on encodings in [23], since they are not available as 0-1 ILP.

Instance Name	Symmetry Statistics			Our Encoding				Encoding in [23]	
	Symm.	Gen.	Saucy Time	W. SBPs		W/o. SBPs		W. SBPs	W/o. SBPs
				ZChaff	Pueblo	ZChaff	Pueblo	ZChaff	ZChaff
bibd(7,7,3,3,1)	2.54e7	12	0	0.08	0	0.01	0	0.29	T/O
bibd(6,10,5,3,2)	2.61e9	14	0	0.54	0	0.03	0	54.24	T/O
bibd(7,14,6,3,2)	4.39e14	19	0.01	0.38	0.01	1.25	0.01	T/O	T/O
bibd(9,12,4,3,1)	1.73e14	19	0.02	0.64	0.01	1.89	0.013	T/O	T/O
bibd(8,14,7,4,3)	3.51e15	20	0.02	0.72	0.01	1.57	0	T/O	T/O

are difficult for many SAT solvers, but are solved by CP solvers in a few minutes. These encodings are available at [13], with and without symmetry-breaking clauses from [23]. Table 1 shows a comparison of both encodings. The table shows instance parameters, followed by Saucy statistics, ZChaff and Pueblo runtimes for our encoding, and ZChaff runtimes for encodings from [23] with and without SBPs. Pueblo does not accept instances without 0-1 ILP constraints. Both Pueblo and ZChaff solve all instances with our encoding in a few seconds, but ZChaff times out on several instances from [23]. All instances possess symmetries, but Saucy runtimes are negligible.

Social Golfers (SG). This problem seeks to divide $g \times s$ golfers into g groups of size s for each of w weeks. Each golfer must play once a week. Any two golfers play in the same group at most once. An instance is described by its parameters (g, s, w) and named sg(g,s,w) in the results tables. We use the 3-D matrix model and instances from [14]. Instances are tested on ZChaff and Pueblo with and without SBPs.

Saucy runtimes and CNF and 0-1 ILP instances sizes with and without SBPs are shown in Table 2. Runtimes for ZChaff, Pueblo, and Solver (from [14][1]) are shown in Table 3, with best runtimes for an instance in boldface. For this problem, adding SBPs speeds up Pueblo considerably on *unsatisfiable* benchmarks. For *all* cases where Pueblo is slower with SBPs, the instance is satisfiable. ZChaff is faster with SBPs for both SAT and UNSAT cases, but is not competitive with Pueblo. All instances possess large numbers of symmetries. Pueblo is usually competitive with Solver results from [14] on SAT instances without the addition of SBPs. However, on UNSAT instances, SBPs are needed to make it competitive, and are effective in doing so. For the larger instances, Saucy runtimes are significant. This increases the overall time for our flow. However, [14] requires SBPs to be designed and implemented separately for individual problems. Our system is automated and generalized. Moreover, [14] reports results for four models of SBPs: two basic models that assign values to a subset of the variables in an instance (thus forcing assignments that satisfy constraints on the remaining variables), and MinLex and anti-Lex constraints. Here, we report the best results among all models. Given an instance it may not be clear which model to use for best results until several have been tried. There is no model in [14] which consistently performs well for this problem. Our system uses only MinLex SBPs.

Hamming Code Generation (HC). This problem seeks to find $b-$bit code words to code n symbols, where the Hamming distance between two symbols is at least d. An instance is specified by the parameters (n, b, d). We use the matrix model from [14], and report results with and without symmetry-breaking in the last four rows of Tables 2 and 3. The instances hc(10,15,9) and hc(12, 20, 12) are unsatisfiable, and the other two are satisfiable. Results for the first two instances are available in [14], the last two are listed as N/A. We observe that symmetry-breaking is useful for both SAT and UNSAT instances, with greater benefit for UNSAT instances. Adding SBPs speeds up ZChaff in all cases, but it is not competitive with Pueblo and Solver. Results reported from [14] are the best out of several combinations of lexicographic and multiset-ordering SBPs. However, several of these combinations are not competitive with our results using Pueblo with SBPs.

---

[1] Results in [14] are on a logarithmic scale, so our numbers are not exact, but all runtimes are rounded *down* for fairness.

Table 2. Saucy symmetry detection statistics and instance sizes for the social golfers and hamming code generation problems, with and without SBPs. For 0-1 ILP instances, number of PB constraints is given in addition to number of CNF clauses and variables. 'K' and 'M' in instance sizes indicate multiples of one thousand and one million.

Instance Params	Saucy Stats Gen.	Saucy Stats Time	Size with SBPs CNF Var.	Size with SBPs CNF Cl.	Size with SBPs 0-1 ILP Var.	Size with SBPs 0-1 ILP Cl.	Size with SBPs 0-1 ILP PB	Size w/o SBPs CNF Var.	Size w/o SBPs CNF Cl.	Size w/o SBPs 0-1 ILP Var.	Size w/o SBPs 0-1 ILP Cl.	Size w/o SBPs 0-1 ILP PB
sg(2,5,4)	16	0.02	6311	33K	1694	1361	141	6139	32K	1522	721	141
sg(2,6,4)	18	0.02	9076	48K	2418	1835	178	8868	46K	2210	1057	178
sg(2,7,4)	20	0.03	12K	65K	3270	2373	219	12041	63894	3026	1457	219
sg(2,8,5)	24	0.07	22K	125K	5320	3761	300	22K	123K	4962	2401	300
sg(3,5,4)	25	0.09	26K	155K	5645	4138	249	26K	152K	5222	2521	249
sg(3,6,4)	28	0.14	37K	221K	8072	5629	321	37K	219K	7562	3673	321
sg(3,7,4)	31	0.21	51K	299K	10K	7336	402	50K	296K	10K	5041	402
sg(4,5,4)	34	0.30	70K	430K	13K	9115	382	69K	426K	12K	6081	382
sg(4,6,5)	42	0.75	134K	837K	23K	15K	556	132K	831K	22K	11K	556
sg(4,7,4)	42	0.79	135K	829K	25K	16K	634	134K	824K	24K	12K	634
sg(4,9,4)	50	1.75	221K	1.35M	42K	25K	950	220K	1.34M	40K	20K	950
sg(5,4,3)	33	0.26	64K	394K	12K	8502	340	64K	391K	11K	5701	340
sg(5,5,4)	43	0.89	145K	911K	25K	16K	540	144K	906K	24K	12K	540
sg(5,7,4)	53	2.79	281K	1.76M	50K	30K	915	279K	1.75M	48K	23K	915
sg(5,8,3)	53	2.3	250K	1.51M	48K	29K	1050	248K	1.51M	47K	23K	1050
sg(6,4,3)	40	0.61	118K	733K	21K	14K	456	117K	729K	20K	9937	456
sg(6,5,3)	46	1.25	182K	1.13M	33K	20K	651	181K	1.12M	31K	15K	651
sg(6,6,3)	52	2.51	260K	1.61M	47K	28K	882	259K	1.60M	46K	22K	882
sg(7,5,3)	54	3.06	301K	1.89M	52K	32K	847	299K	1.88M	50K	24K	847
sg(7,5,5)	68	11.4	551K	3.55M	87K	54K	1015	547K	3.53M	84K	41K	1015
hc(10,15,9)	38	0.07	32K	206K	5842	3762	45	32K	205K	5552	2701	45
hc(10,10,5)	28	0.04	19K	122K	3892	2487	45	19K	121K	3702	1801	45
hc(10,15,8)	38	0.07	32K	206K	5842	3762	45	32K	205K	5552	2701	45
hc(12,20,12)	50	0.19	66K	426K	11K	7023	66	65K	10K	424K	10K	66

Overall, the use of linearity through 0-1 ILP and symmetries by the addition of SBPs – improves performance considerably. For most unsatisfiable instances, the best results are obtained using Pueblo with SBPs added. For satisfiable instances, Pueblo is not improved by SBPs, and in some cases is actually slower. However, ZChaff benefits from SBPs for both SAT and UNSAT instances. This may be because SBPs have greater impact on variable orderings for Pueblo. In most cases Pueblo's results are competitive with results reported for Solver in [14] over a variety of symmetry-breaking ordering constraints. For the cases where Pueblo is faster with SBPs, the average speedup over its performance without SBPs is 83.2, not including timeouts for the no-SBP version. On satisfiable instances, the average slowdown with SBPs is 5.6, but it is much less than that in most cases and there are no timeouts with SBPs. Our system uses academic solvers whose source code and/or binaries are publicly available, but runtimes are comparable with those of Solver, a highly optimized commercial tool.

All results here use problems with matrix models, which frequently possess large numbers of symmetries by construction. While row and column symmetries can be detected manually in a matrix model, our system provides a way to detect and break these symmetries automatically without having to give it any knowledge of the problem semantics. Moreover, it is not restricted to matrix models, and may be used for problems that are likely to have symmetry, but for which matrix models do not exist. It is also applicable in cases where added constraints may disrupt the symmetry in matrix models, e.g. for instances with "customized" requirements. For example, in the social golfer problem, we can add the constraint that certain pairs of golfers must *never* be in the same group. The present matrix model has symmetry along all three dimensions – groups, weeks and golfers. Adding pairwise constraints for specific golfers would leave only partial symmetry between golfers, which poses more effort for manual identification of symmetries. However, with our method added constraints can be analyzed and surviving symmetries detected without any modification. Even if row/column symmetry between certain rows and columns is destroyed, we can still detect symmetries that exist between specific variables in these rows and/or columns automatically. We also hope to identify problems that can be analyzed using our system, but for which matrix models are not applicable.

6   Conclusion

We present an integrated framework for studying and solving a class of CSPs by reduction to SAT and 0-1 ILP. The framework provides for the specification of constraints in a high-level language and automatic compilation into SAT. Specialized methods for SAT have improved considerably over the last 10 years, but these improvements do not necessarily apply to more sophisticated domains because SAT encodings are not always possible and may introduce inefficiencies due to the loss of structure in problem reductions. Our system automatically detects certain types of structure (linearity and symmetries) during compilation and uses them to produce more efficient encodings.

Linearity is preserved through the use of 0-1 ILP, a comparatively more sophisticated problem with specialized solvers that can use leading-edge techniques for SAT solving. We extend earlier work on symmetry-detection in SAT and 0-1 ILP [9,4] to a more general class of CSPs that use non-binary variables and non-linear operations. Symmetries are detected in high-level input by solving the graph automorphism problem on parse trees. MinLex symmetry-breaking predicates (SBPs) from [3] are added to the resulting SAT/0-1 ILP encodings. Other work [14] has focused on symmetry-breaking ordering constraints for known or declared symmetries in generalized CSPs, but we detect and break symmetries automatically. Empirically, we evaluate our system on the balanced incomplete block design (BIBD), social golfers (SG) and Hamming code generation (HC) problems. We detect large numbers of symmetries in all instances, and show that breaking symmetries produces substantial speedups for the 0-1 ILP solver Pueblo [25] on unsatisfiable instances of the SG and HC problems. For CNF reductions, the SAT solver ZChaff [19] exhibits speedups for both satisfiable and unsatisfiable instances when symmetries are broken. Overall, CNF reductions are not competitive with 0-1 ILP reductions. A somewhat surprising observation is that on many satisfiable instances, Pueblo is slowed down by the addition of symmetry-breaking predicates (SBPs). This may be because adding SBPs to satisfiable instances prevents some solutions from being found by Pueblo. More effective SBPs need to be developed for this case. Runtimes for Pueblo with SBPs added are competitive with Solver runtimes reported in [14] on unsatisfiable instances of the SG and HC problems. We also show that our circuit-based CNF encodings for the BIBD problem are more efficient than those proposed in [23]. In general, our system facilitates the comparison of different SAT encodings, since any encoding can be plugged into our framework and automatically tested on several instances. Also, symmetries detected in high-level input can be used by *any* constraints solver, and by methods that add SBPs for declared symmetries [24, 12]. Our framework can be easily extended to include other types of constraints, and to detect additional symmetry such as value symmetry discussed in Section 4. We plan to release code in the public domain to facilitate experimentation with different problems and encodings. At present, information on how to obtain source code, binaries and sample input files for this project is available at [26].

Our current and future work is focused on extending our system to allow more comprehensive coverage of symmetries. We plan to extend our compiler to allow more operations and different types of constraints, and to support more OPL-like [21] syntax. Another direction is the development of efficient SBPs for non-binary variables and of symmetry-breaking constraints that are more effective on satisfiable instances.

References

1. F. A. Aloul, A. Ramani, I. L. Markov, K. A. Sakallah, "Solving Difficult SAT Instances In The Presence of Symmetry", *IEEE Transactions on CAD*, vol. 22(9), pp. 1117-1137, 2003.
2. F. A. Aloul, A. Ramani, I. L. Markov, K. A. Sakallah, "Generic ILP versus Specialized 0-1 ILP: An Update", *in Proceedings of the International Conference on Computer-Aided Design*, pp. 450-457, 2002.
3. F. A. Aloul, I. L. Markov, K. A. Sakallah, "Shatter: Efficient Symmetry-Breaking for Boolean Satisfiability", *in Proc. Intl. Joint Conf. on AI*, pp. 271-282, 2003.
4. F. A. Aloul, A. Ramani, I. L. Markov, K. A. Sakallah, "Symmetry-Breaking for Pseudo-Boolean Formulas", *in Proceedings of the Asia-South Pacific Design Automation Conference*, pp. 884-887, 2004.
5. O. Bailleux, Y. Boufkhad, "Efficient CNF Encoding of Boolean Cardinality Constraints", *Proc. Principles and Practice of Constr. Prog.*, pp. 109-122, 2003.
6. O. Bailleux, Y. Boufkhad, "Full CNF Encoding: The Counting Constraints Case", in *7th Intl. Conf. on Theory and Applications of SAT Testing*, 2004.
7. D. Chai, A. Kuehlmann, "A fast pseudo-boolean constraint solver", *in Proceedings of the Design Automation Conference*, pp.830-835, 2003.
8. C. H Colbourn, J. H. Dinitz, "The CRC Handbook of Combinatorial Designs", CRC Press, 1996.
9. J. Crawford, M. Ginsburg, E. M. Luks, A. Roy, "Symmetry-breaking predicates for search problems", *in Proc. of the Intl. Conf. on Principles of Knowledge Representation and Reasoning*, pp. 148-159, 1996.
10. P. Darga, "SAUCY Man Page", http://vlsicad.eecs.umich.edu/BK/SAUCY/
11. A.M. Frisch, I. Miguel, T. Walsh, "Cgrass: A System for Transforming Constraint Satisfaction Problems", *Jt. Workshop of ERCIM/CologNet area on Constr. Solving and Constr. Logic Prog.*, pp. 23-26, 2002.
12. I. P. Gent, W. Harvey, T. Kelsey, S. Linton, "Generic SBDD using Computational Group Theory", *in Principles and Practice of Constr. Prog.*, pp. 333-347, 2003.
13. I. P. Gent, T. Walsh, B. Selman, CSPLib Problem Library for Constraints; http://www.csplib.org
14. Z. Kiziltan, "Symmetry Breaking Ordering Constraints", *Doctoral Thesis*, Uppsala University, 2004.
15. A.Lal, B. Choueiry, "Dynamic Detection and Exploitation of Value Symmetries for Non-Binary Finite CSPs", *Workshop on Symmetry in CSPs*, 2003.
16. B. McKay, "Practical Graph Isomorphism", *Congressus Numerantium*, vol. 30, pp. 45-87, 1981.

17. F. Margot, "Exploiting Orbits in Symmetric ILP", Mathematical Programming Ser. B 98, pp. 3-21, 2003.
18. P. Meseguer and C. Torras, "Solving strategies for highly symmetric CSPs", *in Proceedings IJCAI*, pp. 400-405, 1999.
19. M. Moskewicz, C. Madigan, Y. Zhao, L. Zhang, S. Malik, "Chaff: Engineering an Efficient SAT Solver", *in Proc. Design Automation Conf.*, pp. 530-535, 2001.
20. G. Nam, F. Aloul, K. Sakallah, R. Rutenbar, "A Comparative Study of Two Boolean Formulations of FPGA Detailed Routing Constraints", *in Proc. of the Intl. Symposium on Physical Design*, pp. 222-227, 2001.
21. P. van Hentenryck, "The OPL Optimization Programming Language", the MIT Press, 1999.
22. The ECL^{i}PS^e Team, "The ECL^{i}PS^e Constraint Logic Programming System": http://www.icparc.ic.ac.uk/eclipse/
23. S. D. Prestwich, "Balanced Incomplete Block Design as Satisfiability", *in 12th Irish Conference on Artificial Intelligence and Cognitive Science*, 2001.
24. J. F. Puget, "Symmetry Breaking Using Stabilizers", *Principles and Practice of Constraints Prog.*, pp. 585-599, 2003.
25. H. Sheini, The Pueblo solver; http://www.eecs.umich.edu/~hsheini/pueblo
26. A. Ramani and I.L. Markov, "GSymEx": Generic Symmetry Extraction for Constraint Programming Problems; http://vlsicad.eecs.umich.edu/BK/GSymEx/
27. ILOG Solver, http://www.ilog.com/products/solver/
28. J. P. Warners, "A Linear-Time Transformation of Linear Inequalities into Conjunctive Normal Form", *in Information Proc. Letters*, vol. 68(2), pp. 63-69, 1998.

Appendix: Compilation into SAT/0-1 ILP

Below, we describe how constraints are translated into CNF and 0-1 ILP. We use a C-like language for high-level constraint specification, and a customized parser that builds a parse tree for the system of constraints. Compilers for SAT and 0-1 ILP walk the parse tree and translate the constraints into CNF/0-1 ILP formulas, which are handed to SAT/0-1 ILP solvers. Solutions are translated back into a form that is meaningful to the original problem. The input language uses C-like syntax to declare variables and specify constraints. Variables are specified as unsigned integers of varying bit sizes, e.g. int1 represents a 1-bit (binary) variable, etc. The mathematical operators allowed are addition (+), subtraction (-) and multiplication (*). Relational operators may be $<=$, $>=$, $==$, and $! =$ (not-equal constraint). *Complement* notation is allowed to express the negative literal for a binary variable ($x1'$ for $x1$). Numeric constants are allowed as coefficients or as the right-hand-side (RHS) value of equations. Division is not presently supported. The compiler also does not support the use of nested parentheses or unary negation but can be easily extended to do so. Support for more sophisticated language constructs, e.g., those used by OPL [21], may be added in the future. An example of constraint declaration in the input language is shown in Figure 1 in Section 3.

To compile into SAT, Boolean "circuits" are instantiated to carry out mathematical operations. An $n-$bit variable is represented by n binary variables in the CNF instance plus a sign bit (to enable subtraction with 2's complement notation). The size of the CNF circuits depends on the operation to be performed. Ripple-carry adders are instantiated for addition operations, and subtraction is performed using 2's complement representation. Both adder and subtractor circuits are linear in the input size. Multiplication is implemented using circuits for Booth's algorithm which are quadratic in the

Table 3. Results for social golfers and Hamming code generation problems. Best results for a given instance are boldfaced. T/O indicates timeout at 600s. The last column shows results from [14]. For UNSAT instances, using Pueblo with SBPs generally performs best. For SAT instances Pueblo is slowed down by SBPs, however ZChaff benefits from SBPs even on SAT instances. All runtimes are in seconds. Results for the last two instances are not shown in [14], so they are listed as N/A.

Instance	Runtime with SBPs		Runtime w/o SBPs		[14]
	ZChaff	Pueblo	ZChaff	Pueblo	Solver
Params	Time	Time	Time	Time	Time
sg(2,5,4)	0.06	**.003**	0.12	0.01	.01
sg(2,6,4)	0.14	**.006**	0.15	0.01	0.1
sg(2,7,4)	0.31	**0.01**	0.14	0.02	5
sg(2,8,5)	1.25	**0.02**	0.89	0.02	30
sg(3,5,4)	2.27	**0.05**	T/O	7.54	0.5
sg(3,6,4)	1.63	**0.09**	T/O	25.7	0.4
sg(3,7,4)	7.7	**0.17**	120	24.8	0.5
sg(4,5,4)	11.5	0.25	T/O	T/O	**0.2**
sg(4,6,5)	T/O	**0.5**	T/O	T/O	2
sg(4,7,4)	T/O	**0.62**	T/O	T/O	5
sg(4,9,4)	T/O	**1.41**	T/O	T/O	2.5
sg(5,4,3)	17.1	0.37	315	**0.07**	0.1
sg(5,5,4)	300	1.3	T/O	1.17	**0.9**
sg(5,7,4)	T/O	**1.8**	T/O	T/O	7
sg(5,8,3)	107	1.76	T/O	T/O	**0.6**
sg(6,4,3)	496	0.86	T/O	**0.47**	0.5
sg(6,5,3)	T/O	1.9	T/O	1.02	**0.6**
sg(6,6,3)	T/O	2.57	T/O	**0.1**	50
sg(7,5,3)	T/O	3.85	T/O	**1.9**	1K
sg(7,5,5)	T/O	59.2	T/O	37	**20**
hc(10,15,9)	93.4	**0.59**	T/O	T/O	7.2
hc(10,10,5)	T/O	22.2	T/O	T/O	**0.4**
hc(10,15,8)	T/O	**275**	T/O	286	N/A
hc(12,20,12)	T/O	**2.77**	T/O	T/O	N/A

input size. Comparison against RHS values uses a linear comparator circuit. There are some built-in optimizations, e.g. smaller circuits for 1-bit addition and subtraction. 1-bit multiplication uses an AND gate. Circuits with a constant as input are partially evaluated. For compilation into 0-1 ILP, linearity is preserved by stating '+' and '-' operations directly as 0-1 ILP constraints. Inequalities ($\leq$, $\geq$, $==$) are also directly expressed in 0-1 ILP, with no need for comparator circuits. Coefficients can be directly written and not multiplied. Multiplication between variables uses CNF clauses, but multiplier outputs can be added/subtracted as part of a linear constraint.

New Structural Decomposition Techniques for Constraint Satisfaction Problems

Yaling Zheng and Berthe Y. Choueiry

Constraint Systems Laboratory,
University of Nebraska-Lincoln
{yzheng,choueiry}@cse.unl.edu

Abstract. We propose four new structural decomposition techniques for Constraint Satisfaction Problems. We compare these four techniques both theoretically and experimentally with hinge decomposition and hypertree decomposition. Our experiments show that one of our techniques offers the best trade-off between the computational cost of the decomposition and the width of the resulting decomposition tree.

1   Introduction

Many important practical problems such as scheduling, resource allocation, and product configuration can be modeled as a Constraint Satisfaction Problem (CSP), which consists of a set of variables, the domains of these variables, and a set of constraints over these variables restricting allowed combinations of values for variables. Although CSPs are in **NP**-complete in general, decomposition techniques borrowed from the area of databases have been used to characterize tractable classes of CSPs [1–4]. The basic principle is to decompose the CSP into sub-problems that are organized in a tree structure. The subproblems are then solved independently, and the solutions are propagated in a backtrack-free manner along the tree [5] to yield a solution to the initial CSP, as described by Dechter and Pearl [1]. We propose new decomposition techniques and position them in the context of the hierarchy specified by Gottlob et al. [4], which unifies main decomposition strategies and compares them in terms of generality. The main techniques are biconnected decomposition (BICOMP) [6], hinge decomposition (HINGE) [2,3], tree clustering (TCLUSTER) [1], hinge decomposition combined with tree clustering (HINGE^{TCLUSTER}) [2], and hypertree decomposition (HYPERTREE) [7]. These techniques can be further characterized by their computational complexity and the width of the tree they generate (which is the size of the largest sub-problem in the tree). Among the above methods, HYPERTREE is the most general and yields trees with the smallest possible width. However, it remains costly in practice even though its complexity is polynomial [8] (see experiments in Section 8). HINGE is a more efficient but less general strategy than HYPERTREE. In this paper, we generalize HINGE into HINGE$^+$, and introduce CUT as a variation of HINGE. Further, we propose a new technique, TRAVERSE, which we combine with CUT to yield a

B. Faltings et al. (Eds.): CSCLP 2004, LNAI 3419, pp. 113–127, 2005.
© Springer-Verlag Berlin Heidelberg 2005

new technique CaT. In summary, HINGE$^+$ generalizes HINGE, and CaT generalizes CUT. We evaluate our new techniques theoretically and empirically on randomly generated hypergraphs. Our experiments show that CaT provides the best trade-off between the width of the generated tree and the computational cost of the decomposition.

This paper is organized as follows. Section 2 reviews the preliminaries of CSPs. Section 3 introduces HINGE$^+$. Section 4 describes CUT, which is a variation of HINGE$^+$. Section 5 introduces a new technique called TRAVERSE. Section 6 combines CUT and TRAVERSE into CaT. Section 7 establishes the formal relationships among these techniques, and also with respect to HINGE and HYPERTREE. Section 8 demonstrates the effectiveness of CaT on randomly generated problems. Finally, Section 9 concludes the paper.

2   Background

A CSP is defined as a tuple $\mathcal{P} = (\mathcal{V}, \mathcal{D}, \mathcal{C})$, where $\mathcal{V}$ is a set of variables, $\mathcal{D}$ is a set of value domains for the variables, and $\mathcal{C}$ is a set of constraints that restrict the acceptable combination of values to variables. Every constraint $C_i \in \mathcal{C}$ is a relation over a set $S_i \subseteq \mathcal{V}$ of variables, and specifies the set of allowed tuples as a subset of the Cartesian product of the domains of S_i. We denote the set of variables involved in constraint C_i by $\text{SCOPE}(C_i)$, and the union of the scopes of a set of constraints $\{C_i\}$ by $\text{VAR}(\{C_i\})$. A solution to the CSP is an assignment of values to all variables such that all the constraints are simultaneously satisfied. The CSP can be represented by its associated constraint hypergraph. The constraint hypergraph of a CSP $\mathcal{P} = (\mathcal{V}, \mathcal{D}, \mathcal{C})$ is given by $\mathcal{H} = (\mathcal{V}, \mathcal{S})$, where $\mathcal{S}$ is a set of hyperedges corresponding to the scopes of the constraints in the CSP. Figure 1 shows the hypergraph $\mathcal{H}_{cg}$ of a CSP with 22 variables and 16 constraints. The primal graph of a constraint hypergraph $\mathcal{H} = (\mathcal{V}, \mathcal{S})$ is a graph $G = (\mathcal{V}, E)$, where E is a set of edges relating any 2 variables that appear in the scope of a constraint in the CSP. Figure 2 shows the primal graph of $\mathcal{H}_{cg}$. Further, we say that a hypergraph is connected when its corresponding primal graph is connected. Each connected component of the primal graph defines a connected component of the hypergraph.

Acyclic CSPs are those CSPs whose associated constraint hypergraph is acyclic. A constraint hypergraph $\mathcal{H}$ is acyclic iff its primal graph G is chordal (i.e., every cycle of length at least 4 has an edge connecting 2 non-adjacent ver-

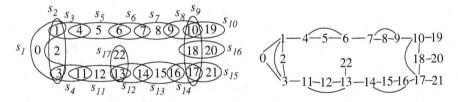

Fig. 1. A constraint hypergraph $\mathcal{H}_{cg}$.   **Fig. 2.** The primal graph of $\mathcal{H}_{cg}$.

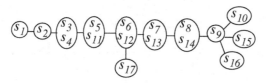

Fig. 3. A join tree of $\mathcal{H}_{cg}$.

tices) and conformal (i.e., there is a one-to-one mapping between each maximal clique of the primal graph and the scope of the constraints) [9]. The constraint hypergraph $\mathcal{H}_{cg}$ shown in Figure 1 is not acyclic.

Following [10], a *join tree* $JT(\mathcal{H})$ for a constraint hypergraph $\mathcal{H}$ is a tree whose nodes are the edges of $\mathcal{H}$ such that whenever the same vertex $X \in \mathcal{V}$ appears in 2 hyperedges s_1 and $s_2 \in \mathcal{S}$, then s_1 and s_2 are connected, and X appears in each node on the unique path linking s_1 and s_2 in $JT(\mathcal{H})$. In other words, the set of nodes in which X appears includes a (connected) subtree of $JT(\mathcal{H})$. The *width* d of a join tree is the maximum number of hyperedges in all the nodes of the join tree. Figure 3 shows a join tree of $\mathcal{H}_{cg}$ of width $d=2$. The principle of *structural decomposition techniques* is to compute an equivalent join tree for a given constraint hypergraph. Each node in this tree is a sub-problem for which we find all solutions, then, while applying directional arc-consistency to the join tree, we can solve the CSP in a backtrack-free manner [1, 2]. The complexity of solving the sub-problems is $O(|\mathcal{S}|l^d d \log l)$, where l is the maximum size of a constraint in $\mathcal{S}$ and d the width of the join tree [2]. Gottlob et al. [4] defined a set of criteria for comparing decomposition methods, where $C(D_i, k)$ is a class of CSPs for which there exists a decomposition of width $\leq k$ by the decomposition method D_i that can be solved in polynomial time. These criteria are as follows (taken verbatim from [4]):

1. *Generalization.* D_2 generalizes D_1 if there exists a constant $\delta \geq 0$ such that, for each level k, $C(D_1, k) \subseteq C(D_2, k+\delta)$ holds. In practical terms, this means that whenever a class C of constraints is tractable according to method D_1, it is also tractable according to D_2.
2. *Beating.* D_2 beats D_1 if there exists an integer k such that $C(D_2, k) \not\subseteq C(D_1, m)$ for any m. Intuitively, this means that some classes of problems are tractable according to D_2 but not according to D_1.
3. *Strong Generalization.* D_2 strongly generalizes D_1 if D_2 generalizes D_1 and D_2 beats D_1. This means that D_2 is really the more powerful method given that, whenever D_1 guarantees polynomial runtime for constraint solving, then D_2 also guarantees tractable constraint solving. However, there are classes of constraints that can be solved in polynomial time by using D_2 but are not tractable according to D_1.
4. *Strongly Incomparable.* D_1 and D_2 are strongly incomparable if both D_1 beats D_2 and D_2 beats D_1.

Figure 4 shows the hierarchy developed by Gottlob et al. [4] based on the above comparision criteria. Whenever two decomposition methods are not related by a directed path, they are strongly incomparable.

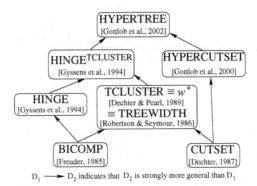

Fig. 4. The hierarchy of constraint tractability of [4].

3   Hinge+ Decomposition (HINGE+)

In this section, we introduce HINGE+ as an improvement of HINGE. As specified by Gyssens et al. [2], HINGE decomposes the constraint hypergraph into a join tree where each node (called 1-hinge) is a set of hyperedges and 2 nodes that are adjacent in the tree share exactly one hyperedge. Figure 5 shows a decomposition of $\mathcal{H}_{cg}$ of Figure 1 by HINGE where $d = 12$. The resulting decomposition guarantees a set of properties (i.e., inheritance, decomposition, and inseparability) that they define. They also attempted to generalize their approach to k-hinges, where a k-hinge is a node in the join tree connected to other nodes with at most k hyperedges. However, they showed that their algorithm for 1-hinge cannot be generalized to achieve a correct result. The width of the join tree of Figure 5 is particularly high. We noticed that by allowing the nodes of the tree to connect through more than 1 hyperedge (as suggested by k-hinge of Jeavons et al. [3]), we can obtain a finer decomposition such as the one shown in Figure 6. We introduce 3 important definitions, which we will use to define HINGE+, our improvment on HINGE:

Definition 1. REMAIN-HG$(F, \mathcal{S})$. *Given a connected constraint hypergraph $\mathcal{H} = (\mathcal{V}, \mathcal{S})$ and a set of hyperedges $F \subseteq \mathcal{S}$, we define $\mathcal{H}_r = (\mathcal{V}_r, \mathcal{S}_r)$, denoted REMAIN-HG$(F, \mathcal{S})$, as the remaining constraint hypergraph obtained after removing F from $\mathcal{S}$. More formally: $\mathcal{V}_r = \mathcal{V} \setminus \mathrm{VAR}(F)$ and $\mathcal{S}_r = \bigcup_{h \in \mathcal{S}} h \setminus \mathrm{VAR}(F)$.*

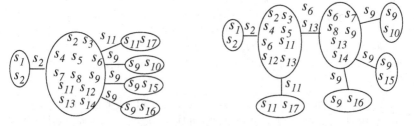

Fig. 5. Applying HINGE to $\mathcal{H}_{cg}$.   **Fig. 6.** A finer decomposition than that of Figure 5.

Definition 2. *i-cut. Given a connected constraint hypergraph $\mathcal{H} = (\mathcal{V}, \mathcal{S})$ where $|\mathcal{S}| \geq i + 1$, an i-cut of $\mathcal{H}$ is a set of hyperedges F such that:*

1. *$F \subset \mathcal{S}$ and $|F| = i$; and*
2. *REMAIN-HG$(F, \mathcal{S})$ has at least 2 components.*

Definition 3. *MAX-SIZE$(F, \mathcal{H})$. Given an i-cut F of a constraint hypergraph $\mathcal{H} = (\mathcal{V}, \mathcal{S})$, MAX-SIZE$(F, \mathcal{H})$ is the largest number of hyperedges in a connected component in REMAIN-HG$(F, \mathcal{H})$.*

Given a constraint hypergraph $\mathcal{H}$, HINGE continuously finds 1-cuts (connecting 1-hinges). We improve HINGE by finding 1-cuts through k-cuts, where k is a specified maximum cut-size. The difficulty here is to choose among the i-cuts for a given i ($1 < i \leq k$), as there may be more than one possible choice. We solve this problem by choosing the i-cut that yields the minimum value of MAX-SIZE. Now we define the join tree resulting from HINGE$^+$:

Definition 4. *k-hinge$^+$-tree. Given a constraint hypergraph $\mathcal{H} = (\mathcal{V}, \mathcal{S})$, a k-hinge$^+$-tree of $\mathcal{H}$ is a tree, $T = (N, A)$, with nodes N and labeled arcs A, such that:*

1. *For each tree node, $p \subseteq \mathcal{S}$;*
2. *For each hyperedge $h \in \mathcal{S}$, there exists a tree node p such that $h \in p$;*
3. *For 2 adjacent tree nodes p_1 and p_2, there exists an i-cut C ($1 \leq i \leq k$) such that VAR$(p_1) \cap$ VAR$(p_2) =$ VAR(C); and*
4. *For each variable $Y \in \mathcal{V}$, the set $\{p \in N \mid Y \in$ VAR$(p)\}$ induces a connected subtree of T.*

Given a constraint hypergraph $\mathcal{H}$ and a constant number k, which is the maximum cut size, HINGE$^+$ (see Algorithm 1) returns a k-hinge$^+$-tree by finding 1-cuts through k-cuts. The worst case of the algorithm occurs when there are no i-cuts $1 \leq i \leq (k-1)$. In this case, line 11 loops at most $|\mathcal{S}|^k$ times, and each loop can be performed in $O(|\mathcal{V}||\mathcal{S}|)$ time. Therefore, the worst-case time complexity of HINGE$^+$ is $O(|\mathcal{V}||\mathcal{S}|^{k+1})$. Since k is used to limit the cut size, Algorithm 1 remains polynomial. Figure 7 shows a 2-hinge$^+$-tree for $\mathcal{H}_{cg}$.

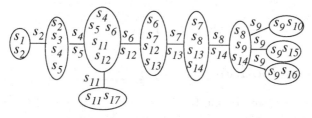

Fig. 7. Applying HINGE$^+$ to $\mathcal{H}_{cg}$ with $k = 2$.

Input: A hypergraph $\mathcal{H} = (\mathcal{V}, \mathcal{S})$ and a maximum cut-size k.

Output: An k-hinge$^+$-tree T for $(\mathcal{V}, \mathcal{S})$.

```
 1  i ← 1;
 2  S_cuts ← ∅;
 3  N_i ← {S};
 4  Mark every hyperedge in S as 'unchosen';
 5  foreach j from 1 to k step by 1 do
 6  │   Mark the nodes in N_i as j-non-minimal;
 7  │   while not all nodes of N_i are marked j-minimal do
 8  │   │   Choose a j-non-minimal node F in N_i;
 9  │   │   j-combinations ← all combinations of j 'unchosen' hyperedges in F;
10  │   │   j-cuts ← ∅;
11  │   │   foreach j-combination X ∈ j-combinations do
12  │   │   │   Γ ← {G ∪ X | G is a connected component in REMAIN-HG(X, F)};
13  │   │   │   if (|Γ| > 1) and (∀C_q ∈ {S_cut | (S_cut ∈ S_cuts) and (S_cut ⊆ F)},
    │   │   │       ∃Γ_p ∈ Γ such that C_q ⊆ Γ_p) then
14  │   │   │   │   j-cuts ← j-cuts ∪ {X};
    │   │   │   end
    │   │   end
15  │   │   if j-cuts ≠ ∅ then
16  │   │   │   choose a j-cut C with smallest MAX-SIZE(j-cut, F);
17  │   │   │   Mark the hyperedges in C as 'chosen';
18  │   │   │   S_cuts ← S_cuts ∪ {C};
19  │   │   │   Γ ← {G ∪ C | G is a connected component in REMAIN-HG(C, F)};
20  │   │   │   N_{i+1} ← (N_i \ {F}) ∪ Γ;
21  │   │   │   Mark C as a j-cut of every element in Γ;
22  │   │   │   Let γ: {FN_1, ..., FN_q} → Γ such that ∀FN_i ∩ γ(FN_i) ≠ ∅;
23  │   │   │   A_{i+1} ←   (A_i \ {({F, F'}, C) | ({F, F'}, C) ∈ A_i})
    │   │   │       ∪{({γ(FN), FN}, C) | ({F, FN}, C) ∈ A_i}
    │   │   │       ∪{({Γ_0, Γ_y}, C) | Γ_0 is an arbitrary chosen element from Γ,
    │   │   │               Γ_y ∈ Γ and Γ_y ≠ Γ_0};
24  │   │   │   Mark all the new nodes added to N_{i+1} as j-non-minimal;
    │   │   else
25  │   │   │   Mark F as j-minimal;
    │   │   end
26  │   │   i ← i + 1;
    │   end
    end
27  T ← (N_i, A_i);
```

Algorithm 1: HINGE$^+$.

4   Cut Decomposition (CUT)

In this section, we introduce CUT as a variation of HINGE$^+$. The arcs incident to every node in the equivalent join tree of a constraint hypergraph obtained by CUT are labeled by at most 2 distinct cuts. For HINGE$^+$, the arcs incident to a given node in an equivalent join tree of a constraint hypergraph obtained by HINGE$^+$ can be labeled by more than 2 distinct cuts. For example, in the join tree of Figure 7, the arcs incident to the node $\{s_4, s_5, s_6, s_{11}, s_{12}\}$ are labeled with three different cuts, namely $\{s_4, s_5\}$, $\{s_6, s_{12}\}$, and $\{s_{11}\}$. The algorithm of CUT is obtained by replacing the conditions in line 13 with the following ones:

1. $|\Gamma| > 1$;
2. For $\forall C_q \in \{S_{\mathrm{cut}} \mid (S_{cut} \in S_{\mathrm{cuts}}) \text{ and } (S_{\mathrm{cut}} \subseteq F)\}$, there exists $\Gamma_p \in \Gamma$ such that $C_q \subseteq \Gamma_p$; and
3. For every 2 sets of hyperedges C_i and $C_j \in S_{\mathrm{cuts}}$, if $C_i \neq C_j$, and $C_i \subseteq \Gamma_i, C_j \subseteq \Gamma_j$, then $\Gamma_i \neq \Gamma_j$.

The above conditions guarantee that no more than 2 cuts label the arcs incident to a node in the join tree obtained by CUT. (This feature allows us to further traverse each tree node from one cut to another cut and is exploited in Section 5.) The complexity of CUT is the same as that of HINGE$^+$. Figure 8 shows the result of applying CUT (the maximum cut size k is 2) to the constraint hypergraph $\mathcal{H}_{cg}$ shown in Figure 1.

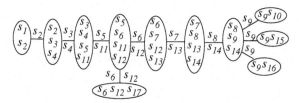

Fig. 8. Applying CUT to $\mathcal{H}_{cg}$.

5   Traverse Decomposition (TRAVERSE)

In this section, we introduce a simple sweep-like decomposition technique called TRAVERSE. We describe two variations of TRAVERSE: TRAVERSE-I and TRAVERSE-II. TRAVERSE-I takes a constraint hypergraph and one set of hyperedges in it, and 'sweeps' through the hypergraph from the set of hyperedges to generate an equivalent join tree of the constraint hypergraph. TRAVERSE-II takes a constraint hypergraph and 2 sets of hyperedges from the hypergraph and 'sweeps' through the constraint hypergraph from the first set of hyperedges to the second set of hyperedges to generate an equivalent join tree of the constraint hypergraph. For convenience, we first introduce the definition of NEIGHBORS(F, S) that will be used in Algorithm 2 and Algorithm 3.

Definition 5. *Neighboring hyperedges. The neighboring hyperedges of a set of hyperedges F in a constraint hypergraph $\mathcal{H} = (V, S)$ with $F \subseteq S$, denoted* NEIGHBORS(F, S), *is a set given by:*

$$\{e \mid e \in F, e \not\subseteq F, \text{ and } \mathrm{VAR}(\{e\}) \cap \mathrm{VAR}(F) \neq \emptyset\}. \tag{1}$$

Given a constraint hypergraph $\mathcal{H} = (V, S)$ and a set of hyperedges $F \subseteq S$, TRAVERSE-I returns a unique join tree obtained by Algorithm 2 via 'sweeping' through the constraint hypergraph starting from the hyperedges in F. We

Input: a constraint hypergraph $\mathcal{H} = (\mathcal{V}, \mathcal{S})$ and a set of hyperedges $F \subseteq \mathcal{S}$.

Output: an equivalent join tree T for $\mathcal{H}$.

1 $N \leftarrow \emptyset; \quad A \leftarrow \emptyset$;
2 Mark any hyperedge $e \in \mathcal{S}$ as '*unvisited*';
3 $F_v \leftarrow \{e \mid \text{VAR}(\{e\}) \subseteq \text{VAR}(F)\}$;
4 $N \leftarrow N \cup \{F_v\}$;
5 $F_{jv} \leftarrow F_v$;
6 Mark any hyperedge in F_{jv} as '*visited*';
7 **while** *not all hyperedges in $\mathcal{S}$ are* 'visited' **do**
8 $\quad$ $F' \leftarrow \text{NEIGHBORS}(F_{jv}, \text{the set of all '\textit{unvisited}' hyperedges})$;
9 $\quad$ $F_v \leftarrow \{e \mid \text{VAR}(e) \subseteq \text{VAR}(F')\}$;
10 $\quad$ $N \leftarrow N \cup \{F_v\}$;
11 $\quad$ $A \leftarrow A \cup \{(F_{jv}, F_v)\}$;
12 $\quad$ $F_{jv} \leftarrow F_v$;
13 $\quad$ Mark every hyperedge in F_{jv} as '*visited*';
end
$T \leftarrow (N, A)$;   **Algorithm 2:** TRAVERSE-I.

denote TRAVERSE-I($\mathcal{H}, F$) the result obtained by applying Algorithm 2 with F on $\mathcal{H}$. The loop in line 7 of Algorithm 2 executes at most $|\mathcal{S}|$ times, and each execution can be performed in $O(|\mathcal{V}||\mathcal{S}|)$ time. Therefore, the worst-case time complexity of TRAVERSE-I is $O(|\mathcal{V}||\mathcal{S}|^2)$. Figure 9 shows the join tree computed by TRAVERSE-I starting from $\{s_1\}$ in $\mathcal{H}_{cg}$. Because it 'sweeps' through the constraint hypergraph, TRAVERSE always computes a join tree that is a connected *chain*, provided the constraint hypergraph is connected. The result of the decomposition depends on F, the starting set of hyperedges. If we traverse $\mathcal{H}_{cg}$ of Figure 1 starting from $\{s_6, s_9, s_{12}\}$, Algorithm 2 would yield a join tree of width $d = 10$. Starting from $\{s_1\}$, the width is $d = 3$ (see Figure 9).

Our goal is to combine CUT with TRAVERSE to improve the k-hinge$^+$-tree computed by CUT (Section 6). To this end, we introduce TRAVERSE-II (Algorithm 3), which allows us to sweep the constraint hypergraph between 2 cuts. TRAVERSE-II takes a constraint hypergraph and 2 sets of hyperedges, and then sweeps through the constraint hypergraph from the first set of hyperedges to the second set of hyperedges to generate an equivalent join tree of this constraint hypergraph. We denote TRAVERSE-II($\mathcal{H}, C_1, C_2$) the result of applying TRAVERSE-II to $\mathcal{H}$ from C_1 to C_2. Figure 10 shows the join tree obtained by applying TRAVERSE-II to $\mathcal{H}_{cg}$ from $\{s_1\}$ to $\{s_9, s_{16}\}$. The loop in line 7 of Algorithm 3 executes at most $|\mathcal{S}|$ times, and each iteration can be performed in $O(|\mathcal{V}||\mathcal{S}|)$ time. Therefore, the complexity of TRAVERSE-II is $O(|\mathcal{V}||\mathcal{S}|^2)$.

Fig. 9. Applying TRAVERSE-I to $\mathcal{H}_{cg}$ from $\{s_1\}$.   **Fig. 10.** Applying TRAVERSE-II to $\mathcal{H}_{cg}$ from $\{s_1\}$ to $\{s_9, s_{16}\}$.

Input: a constraint hypergraph $\mathcal{H} = (\mathcal{V}, \mathcal{S})$, a set of hyperedges C_1 and another set of hyperedges C_2.

Output: an equivalent join tree T for $\mathcal{H}$.

1   $N \leftarrow \emptyset$; $A \leftarrow \emptyset$;
2   Mark any hyperedge $e \in \mathcal{S}$ as '*unvisited*';
3   $F_d \leftarrow \{e \mid \mathrm{VAR}(e) \subseteq \mathrm{VAR}(C_2)\}$;
4   $F_v \leftarrow \{e \mid \mathrm{VAR}(e) \subseteq \mathrm{VAR}(C_1)\}$;
5   $N \leftarrow N \cup \{F_v\}$;
6   Mark any hyperedge in F_{jv} as '*visited*';
7   **while** *($F_v \neq F_d$) and (not all hyperedges in $\mathcal{S}$ are '*visited*')* **do**
8 | $F' \leftarrow \mathrm{NEIGHBORS}(F_{jv} \setminus F_d$, the set of all '*unvisited*' hyperedges $\cup F_d$);
9 | $F_v \leftarrow \{e \mid \mathrm{VAR}(e) \subseteq \mathrm{VAR}(F')\}$;
10 | $N \leftarrow N \cup \{F_v\}$;
11 | $A \leftarrow A \cup \{(F_{jv}, F_v)\}$;
12 | $F_{jv} \leftarrow F_v$;
13 | Mark every hyperedge in F_{jv} as '*visited*';
   end
   $T \leftarrow (N, A)$;

Algorithm 3: TRAVERSE-II.

6   Cut-and-Traverse Decomposition (CaT)

In this section, we introduce CaT, which combines CUT with TRAVERSE. The algorithm of CaT is given in Algorithm 4. Given a constraint hypergraph $\mathcal{H} = (\mathcal{V}, \mathcal{S})$ and a maximum cut size k, Algorithm 4 first applies CUT to $\mathcal{H}$ and generates a k-hinge$^+$-tree in which the arcs incident to any tree node are labeled with at most 2 cuts. This step can be implemented in $O(|\mathcal{V}||\mathcal{S}|^{k+1})$ time. Then, Algorithm 4 applies either TRAVERSE-I or TRAVERSE-II to every tree node in the k-hinge$^+$-tree and generates a set of sub-join trees. Finally, the algorithm combines these sub-join trees into 1 join tree. The traverse process can be performed in $O(|V||\mathcal{S}|^2)$ time. Therefore, the complexity of CaT is $O(|\mathcal{V}||\mathcal{S}|^{k+1} + |V||\mathcal{S}|^2)$. Since $k \geq 1$, the complexity of CaT is $O(|\mathcal{V}||\mathcal{S}|^{k+1})$.

Note that the HYPERTREE algorithm computes an optimal hypertree of $\mathcal{H}$ that has a width within a given bound d; the algorithm returns *failure* if no such decomposition exists [10]. In CaT, the constant k restricts the maximum cut size but does not restrict the width of the generated join tree. Figure 11 and Figure 12 show the equivalent join trees of $\mathcal{H}_{cg}$ computed by CaT and HYPERTREE. In this case, the widths of the join trees obtained by CaT and HYPERTREE are both equal to 2.

7   Characterization

In this section, we compare our 4 techniques with HINGE and HYPERTREE in terms of the criteria proposed by Gottlob et al. [4]. Then, we integrate our results into their hierarchy shown in Figure 4. Finally, we summarize the complexity of all six techniques.

Input: A hypergraph $\mathcal{H} = (\mathcal{V}, \mathcal{S})$ and a maximum cut-size k.
Output: An equivalent join tree T for $\mathcal{H}$.
Cut $\mathcal{H}$ into a tree with tree nodes $P_1, \ldots, P_m$ by CUT;
$N \leftarrow \emptyset; \quad A \leftarrow \emptyset;$
foreach i *from 1 to m* **do**
    switch *the number of cuts labeling the arcs incident to P_i;*
    do
        case *0*
        | $(N_i, A_i) \leftarrow$ TRAVERSE-I$(P_i$, any hyperedge in $P_i)$

        case *1*
            /* C is the only cut labeling the arc incident to P_i */
            $(N_i, A_i) \leftarrow$ TRAVERSE-I(P_i, C)

        case *2*
            /* C_1 and C_2 are the cuts labeling the arcs incident to P_i */
            if *the width of* TRAVERSE-II$(P_i, C_1, C_2) \leq$ *the width of*
            TRAVERSE-II(P_i, C_2, C_1) **then**
            | $(N_i, A_i) \leftarrow$ TRAVERSE-II(P_i, C_1, C_2)
            else
            | $(N_i, A_i) \leftarrow$ TRAVERSE-II(P_i, C_2, C_1)
            end
    end
    $N \leftarrow N \cup \{N_i\};$
    $A \leftarrow A \cup \{A_i\};$
end
$T \leftarrow (N, A);$

Algorithm 4: CaT.

First, we introduce two special classes of constraint hypergraphs borrowed from [4]: Circle(n) (see Figure 13) and book(n) (see Figure 14). These graphs are defined as follows. For any $n \geq 3$, *Circle(n)* is a constraint hypergraph having n hyperedges $\{h_1, \ldots, h_n\}$ such that: $h_i = \{X_i, X_{i+1}\}$ for $\forall 1 \leq i \leq n-1$ and $h_n = \{X_n, X_1\}$. For any $n > 0$, *book(n)* is a constraint hypergraph with $2n + 2$ vertices and $3n + 1$ hyperedges that form n squares (pages of the book) with exactly one common edge $\{X, Y\}$. The hyperedges are defined as follows:

- $b_0 = \{X, Y\};$
- $b_{3i+1} = \{X, X_i\}$ for $\forall 1 \leq i \leq n;$
- $b_{3i+2} = \{X_i, Y_i\}$ for $\forall 1 \leq i \leq n;$ and
- $b_{3i+3} = \{Y_i, Y\}$ for $\forall 1 \leq i \leq n.$

Theorem 1. HINGE$^+$ *strongly generalizes* HINGE.

Proof. (HINGE$^+$ beats HINGE.) Consider the graph *Circle(n)* for some $n \geq 3$. It is easy to see that the HINGE width of *Circle(n)* is n, while its HINGE$^+$ width (with a maximum cut size of 2) is no greater than 4. Hence, $\bigcup_{n \geq 3} \{Circle(n)\} \subseteq C(\text{HINGE}^+, 4)$, while $\bigcup_{n \geq 3} \{Circle(n)\} \not\subseteq C(\text{HINGE}, k)$ holds for every $k > 0$.

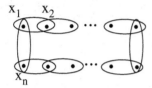

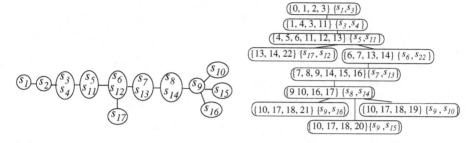

Fig. 11. Applying CaT to $\mathcal{H}_{cg}$.   **Fig. 12.** Applying HYPERTREE to $\mathcal{H}_{cg}$.

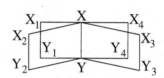

Fig. 13. Circle(n).   **Fig. 14.** Book(4).

Therefore, HINGE$^+$ beats HINGE. (HINGE$^+$ generalizes HINGE.) It is easy to see that HINGE is a special case of HINGE$^+$ when the maximum cut size is 1. Thus, for $\forall I \subseteq C(\text{HINGE}, k)$, $I \subseteq C(\text{HINGE}^+, k)$ holds.   □

Theorem 2. HYPERTREE *generalizes* HINGE$^+$.

Proof. It is obvious that $\forall I \subseteq C(\text{HINGE}^+, k)$, $I \subseteq C(\text{HYPERTREE}, k)$ holds.
    □

Theorem 3. CaT *generalizes* CUT.

Proof. The first phase of CaT is CUT. The second phase of CaT further decomposes each tree node of the join tree obtained by CUT. It is easy to see that $\forall I \subseteq C(\text{CUT}, k)$, $I \subseteq C(\text{CaT, k})$ holds.   □

Theorem 4. HYPERTREE *generalizes* CaT.

Proof. It is obvious that $\forall I \subseteq C(\text{CaT}, k)$, $I \subseteq C(\text{HYPERTREE}, k)$ holds.   □

Theorem 5. HYPERTREE *strongly generalizes* TRAVERSE.

Proof. (HYPERTREE generalizes TRAVERSE.) It is obvious that $\forall I \subseteq C(\text{TRAVERSE}, k)$, $I \subseteq C(\text{HYPERTREE}, k)$ holds. (HYPERTREE beats TRA-VERSE.) Consider the graph *book(n)* for some $n \geq 1$, it is easy to see that the TRAVERSE width of *book(n)* is greater than $\lceil \frac{n}{2} \rceil$, while its HYPERTREE width is 2. Hence, $\bigcup_{n \geq 1} \{book(n)\} \subseteq C(\text{HYPERTREE}, 2)$, while $\bigcup_{n \geq 1} \{book(n)\}$ $\not\subseteq C(\text{TRAVERSE}, k)$ for every $k > 0$.   □

Theorem 6. HINGE *and* TRAVERSE *are strongly incomparable.*

Proof. (HINGE beats TRAVERSE.) Consider the graph *book(n)* for some $n \geq 1$, it is easy to see that the TRAVERSE width of *book(n)* is greater than

$\lceil \frac{n}{2} \rceil$, while its HINGE width is 4. Hence, $\bigcup_{n \geq 1} \{book(n)\} \subseteq C(\text{HINGE}^+, 4)$, while $\bigcup_{n \geq 1} \{book(n)\} \not\subseteq C(\text{HINGE}, k)$ for every $k > 0$. (TRAVERSE beats HINGE.) Consider the graph $Circle(n)$ for some $n \geq 3$. It is easy to see that the HINGE width of $Circle(n)$ is n while its TRAVERSE width (from an arbitrary chosen hyperedge) is 2. Hence, $\bigcup_{n \geq 3} \{Circle(n)\} \subseteq C(\text{TRAVERSE}, 2)$, while $\bigcup_{n \geq 3} \{Circle(n)\} \not\subseteq C(\text{HINGE}, k)$ holds for every $k > 0$. Therefore, TRAVERSE beats HINGE.   □

Theorem 7. CUT *beats* TRAVERSE.

Proof. Consider the graph $book(n)$ for some $n \geq 1$, it is easy to see that the TRAVERSE width of $book(n)$ is greater than $\lceil \frac{n}{2} \rceil$, while its CUT width is 4. Hence, $\bigcup_{n \geq 1} \{book(n)\} \subseteq C(\text{CUT}, 4)$, while $\bigcup_{n \geq 1} \{book(n)\} \not\subseteq C(\text{TRAVERSE}, k)$ for every $k > 0$.   □

Theorem 8. CaT *beats* TRAVERSE.

Proof. Consider the graph $book(n)$ for some $n \geq 1$, It is easy to see that the TRAVERSE width of $book(n)$ is greater than $\lceil \frac{n}{2} \rceil$ while its CaT width (with the maximum cut size being 2) is 2. Hence, $\bigcup_{n \geq 1} \{book(n)\} \subseteq C(\text{CaT}, 2)$, while $\bigcup_{n \geq 1} \{book(n)\} \not\subseteq C(\text{TRAVERSE}, k)$ for every $k > 0$.   □

Theorem 9. HINGE$^+$ *beats* TRAVERSE.

Proof. Consider the graph $book(n)$ for some $n \geq 1$, it is easy to see that the TRAVERSE width of $book(n)$ is greater than $\lceil \frac{n}{2} \rceil$, while its HINGE$^+$ width is 4. Hence, $\bigcup_{n \geq 1} \{book(n)\} \subseteq C(\text{HINGE}^+, 4)$, while $\bigcup_{n \geq 1} \{book(n)\} \not\subseteq C(\text{TRAVERSE}, k)$ for every $k > 0$.   □

Theorem 10. CUT *beats* HINGE.

Proof. Consider the graph $Circle(n)$ for some $n \geq 3$. It is easy to see that the HINGE width of $Circle(n)$ is n, while its CUT width (with maximum cut size being 2) is 2. Hence, $\bigcup_{n \geq 3} \{Circle(n)\} \subseteq C(\text{CUT}, 2)$, while $\bigcup_{n \geq 3} \{Circle(n)\} \not\subseteq C(\text{HINGE}, k)$ holds for every $k > 0$. Therefore, CUT beats HINGE.   □

The above theorems implied that CaT beats HINGE and HYPERTREE generalizes CUT. The relationships between HINGE$^+$ and CUT and between HINGE$^+$ and CaT are still need to be investigated. Figure 15 summarizes the main relationships studied above. The solid directed edge from D_1 to D_2 indicates that D_2 strongly generalizes D_1. The dotted directed edge from D_1 to D_2 indicates D_2 generalizes D_1. Note that the picture is incomplete. Table 1 summarizes the complexity of the techniques shown in Figure 15.

8   Preliminary Experiments

In order to assess empirically the above techniques, we compared their performance on randomly generated hypergraphs in terms of two criteria: the CPU time for computing the decompositions and the width of the resulting join tree.

Table 1. Complexity of decomposition methods.

Technique	Complexity
HYPERTREE	
Normal form: *opt-d-decomp* [7]	$O(|\mathcal{S}|^{2d}|\mathcal{V}|^2)$
Reduced normal form [8]	Best case: $O(|\mathcal{S}|^d|\mathcal{V}| + |\mathcal{S}|^2|\mathcal{V}|)$
HINGE	$O(|\mathcal{V}||\mathcal{S}|^2)$
HINGE$^+$	$O(|\mathcal{V}||\mathcal{S}|^{k+1})$
CUT	$O(|\mathcal{V}||\mathcal{S}|^{k+1})$
TRAVERSE	$O(|\mathcal{V}||\mathcal{S}|^2)$
CaT	$O(|\mathcal{V}||\mathcal{S}|^{k+1})$
Solving the CSP after decomposition	$O(|\mathcal{S}|l^d d \log l)$

$|\mathcal{V}|$: number of variables (i.e., vertices).
$|\mathcal{S}|$: number of constraints (i.e., hyperedges).
d: width of the join tree resulting from a decomposition.
k: maximum cut-size.
l: maximum size of a constraint in $\mathcal{S}$.

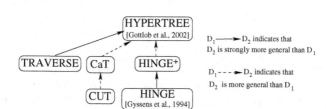

Fig. 15. Illustrating the relationships between the various studied techniques.

For HYPERTREE, we used the algorithm of Harvey and Ghose [8], which improves on the opt-k-decomp algorithm of Gottlob et al. [10]. By starting with k=1 and incrementing its value by 1 until it finds decomposition, the algorithm we used guarantees an optimal decomposition. We generated random hypergraphs setting the number of constraints to 10, 11, 12, and 13. In each instance, we chose the arity of the constraints randomly in $\{2, 3, 4\}$. Table 2 summarizes the constraint hypergraphs used in the experiments. We set the maximum cut size k=2 for HINGE$^+$, CUT, and CaT. Figure 16 and Figure 17 show, for a fixed number of constraints, the average CPU times and average widths of the generated join trees. Figure 16 and Figure 17 show the average CPU times and average widths of different decomposition techniques. Table 3 averages these results over all 4000 instances generated.

From these experiments, we have the following observations:

For CPU time,
    TRAVERSE < HINGE < CUT $\approx$ CaT $\approx$ HINGE$^+$ $\ll$ HYPERTREE.

Table 2. Constraint hypergraphs used in the experiments.

# constraints	# variables	# instances
10	{16, 17, ..., 25}	1000 (100 instances for each fixed number of variables)
11	{18, 19, ..., 27}	1000 (100 instances for each fixed number of variables)
12	{20, 21, ..., 29}	1000 (100 instances for each fixed number of variables)
13	{22, 23, ..., 31}	1000 (100 instances for each fixed number of variables)

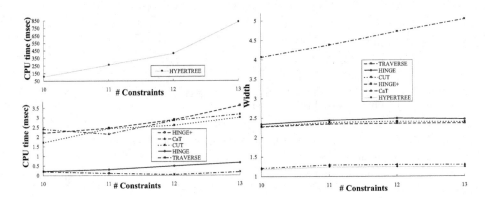

Fig. 16. Average CPU times.   **Fig. 17.** Average widths.

TRAVERSE is the quickest technique followed by HINGE then CaT, HINGE⁺, and CUT, which have comparable values for the CPU time. All techniques are significantly quicker than HYPERTREE. Indeed, the computationally cost of HYPERTREE is prohibitively high although its worst-case time complexity is polynomial.

For width,

HYPERTREE ≈ CaT < HINGE⁺ ≈ CUT ≈ HINGE < TRAVERSE.

The join tree obtained with TRAVERSE has the largest width. The average widths of the join tree generated by HINGE⁺ and CUT are smaller than that of the join tree generated by HINGE. However, the differences of these values are within 4%. The widths of the join trees generated by CaT and HYPERTREE differ by only 4%, which is negligible. Also, they are significantly smaller than those generated by the remaining techniques.

In summary, CaT offers the best trade-off between the CPU time and the width of the computed join tree among the decomposition methods tested.

9   Conclusion

In this paper, we proposed two main new structural decompositions: HINGE⁺ and CaT. HINGE⁺ strongly generalizes HINGE of Gyssens et al. [2]. CaT is built by combining CUT (a variation of HINGE⁺) and TRAVERSE (a sweep-like

Table 3. Average results over all 4000 instances.

Comparison criteria Average	HINGE	HINGE$^+$	CUT	TRAVERSE	CaT	HYPERTREE
CPU time [msec]	0.400	2.786	2.428	0.130	2.640	400.900
Width	2.425	2.332	2.367	4.547	1.273	1.225

decomposition techniques). We compared these techniques among themselves and with HINGE and HYPERTREE both theoretically and experimentally. Our experiments showed that the CaT offers the best trade-off between cost and quality of the resulting decomposition.

In the future, we plan to address the following issues: (1) Compare our techniques with the remaining techniques shown in Figure 4; and (2) Perform experiments on special types of graphs (e.g., small-world graphs and clustered graphs) and real-world problems (e.g., the ones used in [11]).

Acknowledgments

This work is supported by CAREER Award #0133568 from the National Science Foundation. The experiments were conducted utilizing the Research Computing Facility of the University of Nebraska-Lincoln. Deb Derrick provided invaluable editorial help.

References

1. Dechter, R., Pearl, J.: Tree Clustering for Constraint Networks. Artificial Intelligence **38** (1989) 353–366
2. Gyssens, M., Jeavons, P.G., Cohen, D.A.: Decomposing Constraint Satisfaction Problems Using Database Techniques. Artificial Intelligence **66** (1994) 57–89
3. Jeavons, P.G., Cohen, D.A., Gyssens, M.: A Structural Decomposition for Hypergraphs. Contemporary Mathematics **178** (1994) 161–177
4. Gottlob, G., Leone, N., Scarcello, F.: A Comparison of Structural CSP Decomposition Methods. Artificial Intelligence **124** (2000) 243–282
5. Freuder, E.C.: A Sufficient Condition for Backtrack-Free Search. JACM **29 (1)** (1982) 24–32
6. Freuder, E.C.: A Sufficient Condition for Backtrack-Bounded Search. JACM **32 (4)** (1985) 755–761
7. Gottlob, G., Leone, N., Scarcello, F.: Hypertree Decompositions and Tractable Queries. Journal of Computer and System Sciences **64** (2002) 579–627
8. Harvey, P., Ghose, A.: Reducing Redundancy in the Hypertree Decomposition Scheme. In: The 15th IEEE International Conference on Tools with Artificial Intelligence (ICTAI 03). (2003) 474–481
9. Dechter, R.: Constraint Processing. Morgan Kaufmann (2003)
10. Gottlob, G., Leone, N., Scarcello, F.: On Tractable Queries and Constraints. In: 10th International Conference and Workshop on Database and Expert System Applications (DEXA 1999). (1999) 1–15
11. Gottlob, G., Hutle, M., Wotawa, F.: Combining Hypertree, Bicomp, And Hinge Decomposition. In: Proc. of the 15 th ECAI, Lyon, France (2002) 161–165

Algorithms for the Maximum Hamming Distance Problem

Ola Angelsmark[1,*] and Johan Thapper[2,**]

[1] Department of Computer and Information Science,
Linköpings Universitet,
S-581 83 Linköping, Sweden
olaan@ida.liu.se

[2] Department of Mathematics,
Linköpings Universitet,
S-581 83 Linköping, Sweden
jotha@mai.liu.se

Abstract. We study the problem of finding two solutions to a constraint satisfaction problem which differ on the assignment of as many variables as possible – the MAX HAMMING DISTANCE problem for CSPs – a problem which can, among other things, be seen as a domain independent way of quantifying "ignorance." The first algorithm we present is an $\mathcal{O}(1.7338^n)$ microstructure based algorithm for MAX HAMMING DISTANCE 2-SAT, improving the previously best known algorithm for this problem, which has a running time of $\mathcal{O}(1.8409^n)$. We also give algorithms based on enumeration techniques for solving both MAX HAMMING DISTANCE l-SAT, and the general MAX HAMMING DISTANCE (d, l)-CSP, the first non-trivial algorithms for these problems. The main results here are that if we can solve l-SAT in $\mathcal{O}(a^n)$ and (d, l)-CSP in $\mathcal{O}(b^n)$, then the corresponding Max Hamming problems can be solved in $\mathcal{O}((2a)^n)$ and $\mathcal{O}(b^n(1 + b)^n)$, respectively.

1   Introduction

In its most basic form, a constraint satisfaction problem (CSP) consists of a collection of variables taking values from some domain, and a collection of constraints restricting the values different variables can simultaneously assume. The question here is: Can we find an assignment of values to the variables which does not violate any of the constraints? While this is certainly the most thoroughly studied problem for CSPs, there are a number of alternative, equally interesting, questions one can ask about a CSP. The question we will study in this paper asks us to find two solutions that are as far away from each other as possible; i.e. we want to find two satisfying assignments that disagree on the values for as many variables as possible. This is known as the MAX HAMMING DISTANCE

* Supported in part by the National Graduate School in Computer Science, Sweden, and in part by the *Swedish Research Council* (VR), grant 621-2002-4126.
** Supported by the *Programme for Interdisciplinary Mathematics*, Department of Mathematics, Linköpings Universitet.

B. Faltings et al. (Eds.): CSCLP 2004, LNAI 3419, pp. 128–141, 2005.

problem, and was first introduced in Crescenzi & Rossi [3], where it was suggested as a domain independent measure of ignorance, quantifying how much we do not know of the world.

We present three different algorithms. The first one is a microstructure based algorithm for the special case when the domains have size two and we have binary constraints, denoted MAX HAMMING DISTANCE $(2,2)$-CSP. (We will exclusively consider CSPs over finite domains, denoted (d,l)-CSP, where d is the domain size and l the arity of the constraints.) Even in this restricted form, the problem is NP-complete. In the microstructure graph of a CSP [9], a vertex corresponds to an assignment of a value to a variable in the original problem (see Section 2 for definitions.) The algorithm exploits this by searching for a set of vertices where each vertex either does not have an edge to any other vertex – and thus can be interpreted as an assignment – or is part of a connected component with 2 or 4 vertices. Each vertex (i.e. assignment) in this set is then given a weight, and the original instance together with these weights is given to a weighted 2-SAT solver. This algorithm returns a solution with maximum weight W, and we can then construct a solution which differs on W variables.

By using the weighted 2-SAT algorithm from [4], we arrive at a running time of $\mathcal{O}(1.7338^n)$, where n is the number of variables in the problem. This is an improvement over the MAX HAMMING DISTANCE $(2,2)$-CSP algorithm presented in [1], which runs in $\mathcal{O}(1.8409^n)$.

When we allow domains with more than 2 elements, or constraints with arity higher than 2, it turns out that microstructures are not as successful, and, consequently, the algorithms we present for these cases are quite different. Intuitively, the algorithm for MAX HAMMING DISTANCE (d,l)-CSP works as follows:

1. Pick a subset of the variables which should assume different values in the two solutions, duplicate and rename them and the constraints they are involved in.
2. Add a constraint for each of these new variables, preventing it from assuming the same value as the one it is a copy of.
3. Solve this new, larger, instance.

Starting with assuming all variables are different in the two solutions, and then working downwards, the algorithm will, by trying out the different possible subsets of variables, arrive at a pair of solutions with maximum hamming distance. Given that we can solve the (d,l)-CSP in the last step in time $\mathcal{O}(a^n)$, the entire algorithm will have a running time of $\mathcal{O}(a^n(1+a)^n)$.

The final algorithm is for the case when the domain has two elements and the constraints have arity l, MAX HAMMING DISTANCE $(2,l)$-CSP. Here, we note that since there are only two possible choices of values for a variable, it is unnecessary to duplicate the variables that should take different values – instead, only the constraints they are involved in are duplicated, and then any occurrence of a variable which should assume different values in the two solutions is replaced by its negation in these constraints. The resulting algorithm will have a running time of $\mathcal{O}((2a)^n)$, where $\mathcal{O}(a^n)$ is the time needed to solve the $(2,l)$-CSP problem in each step.

Overview of the Paper: Section 2 contains most of the definitions we will need in the discussion. For convenience, it has been split into three parts, where Section 2.1 contains the definitions related to CSPs, Section 2.2 the graph and microstructure definitions, while Section 2.3 formally defines the problem we will be studying, MAX HAMMING DISTANCE. The algorithm for MAX HAMMING DISTANCE $(2,2)$-CSP, together with its analysis, is presented in Section 3, while Section 4 contains the algorithms for MAX HAMMING DISTANCE (d,l)-CSP and MAX HAMMING DISTANCE $(2,l)$-CSP.

2   Preliminaries

This section is divided into three parts in order simplify the search for a particular definition. In Section 2.1 we have the definitions related to constraint satisfaction problems, while Section 2.2 is devoted to graphs and the microstructure of CSPs. Finally, in Section 2.3, we define the problem we will be discussing in this paper; MAX HAMMING DISTANCE (d,l)-CSP. Note that Section 3 contains additional definitions specific to that part of the paper.

2.1   Constraint Satisfaction Problems

A (d,l)-*constraint satisfaction problem* $((d,l)$-*CSP)* is a triple (X, D, C) where

- X is a finite set of variables,
- D a finite set of domain values, with $|D| = d$, and
- C is a set of constraints $\{c_1, c_2, \ldots, c_k\}$.

Each constraint $c_i \in C$ is a structure $R(x_{i_1}, \ldots, x_{i_j})$ where $j \leq l, x_{i_1}, \ldots, x_{i_j} \in X$ and $R \subseteq D^j$. A *solution* to a CSP instances is a function $f : X \to D$ s.t. for each constraint $R(x_{i_1}, \ldots, x_{i_j}) \in C$, $(f(x_{i_1}), \ldots, f(x_{i_j})) \in R$. Given a (d,l)-CSP, the basic computational problem is to decide whether it has a solution or not – to determine if it is *satisfiable*.

The special case when $d = 2$ and we have binary constraints, i.e. $(2,2)$-CSP, will often be viewed as 2-SAT formulae. A 2-SAT formula is a conjunction of a number of clauses, where each clause is on one of the forms $(p \vee q)$, $(\neg p \vee q)$, $(\neg p \vee \neg q)$, (p), $(\neg p)$. The set of variables of a formula F is denoted $\mathrm{Var}(F)$, and an occurrence of a variable or its complement in a formula is termed a *literal*. Determining whether a 2-SAT formula is satisfiable can be done in polynomial time [2], while, in contrast, the more general l-SAT (i.e, the clauses consist of at most l literals) is known to be NP-complete for $l \geq 3$ [7].

Definition 1 ([4]). *Let F be a 2-SAT formula, and let L be the set of all literals for all variables occurring in F. Given a vector $\boldsymbol{w}$ of weights and a model M for F, we define the weight $W(M)$ of M as*

$$W(M) = \sum_{\{l \in L \ | \ l \ is \ true \ in \ M\}} \boldsymbol{w}(l)$$

The problem of finding a maximum weighted model for F is denoted 2-SAT$_w$.

In [4], an algorithm for counting the number of maximum weighted solutions to 2-SAT instances is presented which has a running time of $\mathcal{O}(1.2561^n)$, and it can easily be modified to return one of the solutions.

2.2 Graphs and Microstructures

A *graph* G consists of a set $V(G)$ of *vertices* and a set $E(G)$ of *edges*, where each element of $E(G)$ is an unordered pair of vertices. The *neighbourhood* of a vertex $v \in V(G)$ is the set of all vertices adjacent to v, excluding v itself, and is denoted $N_G(v)$, $N_G(v) := \{u \in V(G) \mid (v, w) \in E(G)\}$. If, by following the edges of the graph, we can get from a vertex v to v', then v' is *reachable* from v. The *connected components* of a graph are the equivalence classes of vertices under the "is reachable from" relation.

Definition 2 ([9]). *Given a binary CSP $\Theta = (X, D, C)$, the microstructure of Θ is an undirected graph G, defined as follows:*

1. *For each variable $x \in X$, and domain value $d \in D$, there is a vertex $x[d]$ in G.*
2. *There is an edge $(x[d], y[e]) \in E(G)$ iff (d, e) satisfies the constraint between x and y.*

We assume that there is exactly one constraint between any pair of variables, and variables with no explicit constraint between them is assumed to be constrained by the universal constraint which allows all values.

For convenience, we will work exclusively with the complement of the graph in Definition 2. The complement of a (microstructure) graph G is a graph containing exactly those edges which are not present in G (excluding loops), i.e. a graph with edge set $\{(v, u) \mid v \neq u \wedge (v, u) \notin E(G)\}$.

A variable with domain size d will in the microstructure graph be a clique of size d. When the domain has two elements and we have a clique of size 2, we let $x[i]$ denote an arbitrary value for x, and use $x[1 - i]$ to denote the other possible value. For example, if we look at the 2-SAT formula $(x \vee y) \wedge (\neg x \vee z)$, with domain values 0 and 1, it has the microstructure graph shown in Fig. 1. One independent set the graph is $\{x[0], y[1], z[0]\}$ which corresponds to the satisfying assignment $\{x \mapsto 0, y \mapsto 1, z \mapsto 0\}$.

2.3 Hamming Distance of CSPs

The algorithms we present in this paper are all designed to solve different variants of the MAX HAMMING DISTANCE problem for constraint satisfaction problems [3]. Since our algorithms are not limited to problems with two valued domains, the following definition differs somewhat from the one given in [3]:

Definition 3. *Given a set of variables X over finite domains, the Hamming distance between a pair f_1 and f_2 of assignments of values to the variables in X, denoted $d_H(f_1, f_2)$, is the number of variables on which f_1 and f_2 disagree.*

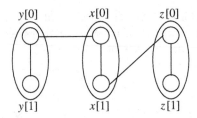

Fig. 1. The microstructure graph of $(x \vee y) \wedge (\neg x \vee z)$.

algorithm $MH1(\alpha, G, \Theta)$

1. **if** $\delta(x) \in \{(3,1), (2,2), (2,1), (1,1)\}$ for all variables x in G **then**
2.   **return** $MH2(\alpha, G, \Theta)$
3. **end if**
4. Choose a variable x in G with $\delta(x) \in \{(\geq 3, \geq 2), (\geq 4, 1)\}$
5. $(\alpha_0, \beta_0) = MH1(\alpha \cup \{x[0]\}, G - N_G(x[0]) - \{x[0]\}, \Theta)$
6. $(\alpha_1, \beta_1) = MH1(\alpha \cup \{x[1]\}, G - N_G(x[1]) - \{x[1]\}, \Theta)$
7. **return** $(\alpha_i, \beta_i), i \in \{0,1\}$ maximising $d_H(\alpha_i, \beta_i)$

Fig. 2. The main algorithm for MAX HAMMING DISTANCE $(2,2)$-CSP.

For example, consider the 2-SAT formula $(x \vee y) \wedge (\neg x \vee z)$, and the two assignments $f_1 = \{x \mapsto 0, y \mapsto 1, z \mapsto 0\}$, $f_2 = \{x \mapsto 1, y \mapsto 1, z \mapsto 1\}$. Clearly, both f_1 and f_2 satisfy the formula, and their Hamming distance is 2, since they disagree on the values for x and z.

The following formalises the problem:

Definition 4 (Maximum Hamming Distance of (d, l)-CSPs). *Let $\Theta = (X, D, C)$ be an instance of (d, l)-CSP. The MAX HAMMING DISTANCE (d, l)-CSP problem is to find two satisfying assignments f and g to Θ which maximises $d_H(f, g)$.*

A naïve enumeration algorithm for this problem would have a time complexity of $\mathcal{O}\left(d^{2n}\right)$. In the following sections we will present ways to significantly improve this running time.

3 Algorithm for Max Hamming Distance $(2, 2)$-CSP

In this section we will discuss and analyse our algorithm for MAX HAMMING DISTANCE $(2,2)$-CSP. Since the formulae for the time complexity of the algorithm can be rather lengthy, the final step, that of calling a weighted 2-SAT solver for every leaf in the search tree, has been left out (unless otherwise noted.) Furthermore, we will omit polynomial factors in the time complexities.

Before we start the discussion of the algorithms, we will need some additional definitions: The *degree* of a vertex v in a graph, usually denoted $\deg(v)$, is the size of its neighbourhood, i.e. $|N_G(v)|$. However, we are not really interested in the degree of a single vertex, but rather in the degrees of the two vertices that

algorithm $MH2(\alpha, G, \Theta)$

1. **if** $\delta(x) \in \{(2,1),(1,1)\}$ for all variables x in G **then**
2.   **return** $MH3(\alpha, G, \Theta)$
3. **end if**
4. **if** G contains a cycle **then**
5.   **if** all variables x has $\delta(x) = (2,2)$ in a cycle **then**
6.   Choose x in this cycle
7.   **else if** there is a variable z with $\delta(z) = (2,2)$ in a cycle **then**
8.   Choose x in a cycle s.t $\delta(x) = (3,1)$ and $x[i]$ has a
               neighbour y with $\delta(y) = (2,2)$
9.   **else**
10.   Choose x with $\delta(x) = (3,1)$ in a cycle
11.   **end if**
12. **else** % G is cycle-free
13.   Choose x which is two variables from the end of a chain, if possible,
          otherwise, choose x one variable from the end of a chain
14. **end if**
15. $(\alpha_0, \beta_0) = MH2(\alpha \cup \{x[0]\}, G - N_G(x[0]) - \{x[0]\}, \Theta)$
16. $(\alpha_1, \beta_1) = MH2(\alpha \cup \{x[1]\}, G - N_G(x[1]) - \{x[1]\}, \Theta)$
17. **return** $(\alpha_i, \beta_i), i \in \{0,1\}$ maximising $d_H(\alpha_i, \beta_i)$

Fig. 3. The helper function *MH2*.

make up a variable. Thus let $\Theta = (X, D, C)$ be a $(2,2)$-CSP and, for $x \in X$, define the *variable degree* $\delta(x)$ as a tuple $(\deg(x[i]), \deg(x[1-i]))$, where $x[i]$ is the vertex with highest degree. If we are interested in variables with degrees higher than a certain number, we write $\delta(x) = (\geq i, \geq j)$.

In the analysis of the algorithm in this section, we will often encounter recursions on the form $T(n) = \sum_{i=0}^{k} T(n - r_i) + p(n)$, where $p(n)$ is a polynomial in n and $r_i \in \mathbb{Z}^+$. These equations satisfy $T(n) \in \mathcal{O}(\tau(r_1, r_2, \ldots, r_k)^n)$, where $\tau(r_1, r_2, \ldots, r_k)$ is the largest real-valued solution to the equation $1 - \sum_{i=1}^{k} x^{-r_i} = 0$ (see Kullman [10].) Note that this bound depends on neither $p(n)$ nor the boundary conditions $T(1) = b_1, \ldots, T(k) = b_k$. We will sometimes refer to τ as the *work factor* (in the sense of [5].)

With that in mind, we are now ready to discuss the algorithm, which consists of three functions: *MH1*, the main algorithm, which calls *MH2* once no variable is involved in more than three constraints, which, in turn, calls *MH3* when every variable is involved in at most one constraint.

MH1: The main algorithm, *MH1*, given in Fig. 2, takes as input a partial assignment α, a microstructure graph G, and the original problem instance Θ. If every variable in the microstructure is involved in less than 3 constraints, the helper function $MH2$ is called. In the graph, this translates to every variable x having $\delta(x)$ in the set $\{(3,1),(2,2),(2,1),(1,1)\}$. Otherwise, a variable involved in more than 3 constraints is chosen, and the algorithm branches on the two possible values. We note that for $\delta(x) = (3,2)$, there will be at least 3 variables less in one branch and 2 variables less in the second branch, and for $\delta(x) = (4,1)$, there are at least 4 and 1 variables less, respectively. Consequently, we get work factors of $\tau(3,2)$ and $\tau(4,1)$ for these cases, where $\tau(4,1)$ clearly dominates.

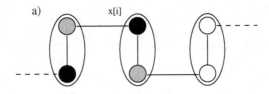

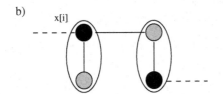

Fig. 4. Branching on $x[i]$ will remove the shaded values and force the black values.

Example 1. Consider the 2-SAT formula $\Theta = (x \lor y) \land (\neg x \lor y)$, which has the microstructure shown in Fig. 1. The variables in Θ have degrees $\delta(x) = (2, 2), \delta(y) = (2, 1)$ and $\delta(z) = (2, 1)$, thus we immediately jump to algorithm *MH2*. Had this not been the case, a variable with a degree of $(\geq 3, \geq 2)$ or $(\geq 4, 1)$ would have been chosen, and the algorithm would have branched on either of the possible values for it.

MH2: The first helper function, *MH2*, shown in Fig. 3, takes over when every variable is involved in zero, one or two constraints. Unless there are no variables involved in two constraints (in which case we jump straight to algorithm *MH3*), we start with checking for cycles. Any cycles we encounter need to be broken, which is done on lines 4 to 11.

First of all, if there is a cycle where every variable has a degree of $(2, 2)$, then selecting one value for a variable in this cycle will propagate through the entire cycle, as is shown in the top part of Fig. 4. On line 8, by choosing a variable x with $\delta(x) = (3, 1)$ with a neighbour y with $\delta(y) = (2, 2)$, one of the values for x will propagate through y. (See Fig. 4b.) Consequently, 4 variables are removed in one branch, and one in the other, giving a work factor of $\tau(4, 1)$ for this case. The obvious exception is when the cycle contains only 3 variables, as is shown in Fig. 7a. Note that the coloured vertex $x[i]$ is the only possible choice – the other assignment would lead to an inconsistency.

Now if every variable x in the cycle has $\delta(x) = (3, 1)$, we get a number of different possibilities, but before we discuss them, we need to make some observations. Once a variable has no neighbours (Fig. 6a), or is part of a component consisting of only two variables and one edge between them, a 'hurdle,' (Fig. 6b), we need no longer consider it. The first case is obvious, since if a variable has no neighbours, it is not involved in any constraints, and we can choose its value freely, while the second case is somewhat harder; We will get back to it when we discuss algorithm *MH3*. Consequently, when a component of $(3, 1)$ variables has

algorithm $MH3(\alpha, G, \Theta)$

1. Let w be a vector of weights, initially all set to 0
2. **for each** $x[i] \in \alpha$ **do**
3.   add weight $w(x[1 - i]) := 1$
4. **for each** connected component of G **do**
5.   Add weights to w, as shown in Fig. 6.
6. $(\beta, W) := 2\text{-}SAT_w(\Theta, w)$
7. **for each** variable x in G **do**
8.   **if** $x[i]$ in β **then**
9.   If possible, add $x[1 - i]$ to α, otherwise, add $x[i]$.
10.   **end if**
11. **end for**
12. **return** (α, β)

Fig. 5. The helper function *MH3*.

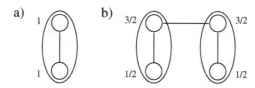

Fig. 6. Variables with no or exactly one neighbour, and the weights they are given by algorithm *MH3*.

at most 3 variables, as in Fig. 7b, when choosing such a variable, in effect, the entire component is removed from the problem and need no longer be considered – in one case we get a unique assignment for the remaining (black) vertices, and in the other case we get a hurdle. If there are no cycles in the component, e.g we have a 'comb-like' structure, as in Fig. 8, then choosing any of the three variables to branch on will, again, remove the entire component, giving a work factor of $\tau(3, 3)$. This also holds for cycle-free components of size 4 and 5. When there are more than 5 variables in the component, by choosing a variable which is two variables removed from the end of the comb (the marked variable in Fig. 8), the chain is broken and we remove 3 variables in one branch and 4 in the other. As was seen in the case for cycles where all variables have degree $(2, 2)$, the number of removed variables increases if a neighbour of the branching variable has this property. Consequently, we will focus on the combs and merely note that the time complexity will not be worse if we have more variables with degree $(2, 2)$.

Getting back to discussing cycles; When we reach line 10 of algorithm *MH2*, every cycle consists exclusively of variables with degree $(3, 1)$, and since no vertex in the graph has degree higher than 3, there can be at most one cycle in a component. The case with cycles containing 3 variables was discussed earlier, and for the case with 4 variables we get one branch where the entire component is removed, and one where we get a comb with 3 variables, which can be removed in its entirety when we branch. There can be no more than $n/4$ cycles with 4 variables in the graph at this point. For each of these cycles, we choose one

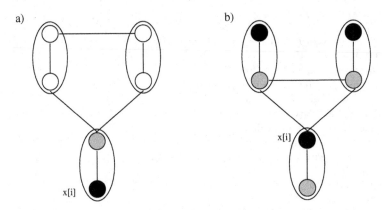

Fig. 7. The case when a component is a cycle with 3 variables.

variable to branch on, and in one branch the entire component is removed, while in the other, we get a component with 3 variables. Since we want to look at all of these cycles, and both branches, this is equivalent to selecting k cycles where we remove the entire component, and then examine the remaining $n/4 - k$ components. In other words, it will require

$$\sum_{k=0}^{n/4} \binom{n/4}{k} \left(1^k \cdot \tau(3,3)^{3(n/4-k)} \right)$$

steps to examine all the cycles. Using the binomial theorem, this can be simplified to $(1 + \tau(3,3)^3)^{n/4}$.

For cycles with 5 variables, the situation is similar, but for 6 we no longer remove the entire component in one of the branches. Instead, we get one branch with 5 variables, and one with 3, which, using the same reasoning as above, gives

$$\sum_{k=0}^{n/6} \binom{n/4}{k} \left(\tau(3,3)^{3k} \cdot \tau(5,5)^{5(n/6-k)} \right) = (\tau(3,3)^3 + \tau(5,5)^5)^{n/6}.$$

Similarly, for cycles of length 7, we get $(\tau(4,4)^4 + \tau(3,4)^6)^{n/7}$. In general, if we have cycles of length c, one branch will have one variable less, and the other three variables less, giving the following general running time:

$$\sum_{k=0}^{n/c} \binom{n/c}{k} \left(\tau(3,4)^{(c-3)k} \cdot \tau(3,4)^{(c-1)(n/c-k)} \right) =$$

$$= \left(\tau(3,4)^{c-1} + \tau(3,4)^{c-3} \right)^{n/c} < (2\tau(3,4)^c)^{n/c} = (2^{1/c}\tau(3,4))^n.$$

Example 2 (cont'd). When we reach algorithm *MH2* with our (still unchanged) microstructure graph, we note that for x, $\delta(x) = (2,2)$, thus we will continue past the test on line 1. Since the graph is cycle-free, we will choose a variable which is not on the end of a chain – in our case the only choice is x, and branch on the two possible values for x.

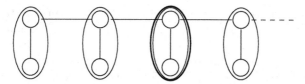

Fig. 8. Choosing a variable in a comb with more than 4 variables.

1. The branch with $x[0]$ removes $y[0]$ and $x[1]$ from the graph, forcing $y[1]$ and leaving z unconstrained.
2. For $x[1]$, $x[0]$ and $z[0]$ are removed, forcing $z[1]$ and leaving y unconstrained.

Consequently, both of these branches will result in *MH3* being called in the next recursive call (since we have a variable degree of $(1,1)$ for the unconstrained variables left in the graph), with $\alpha = \{x[0], y[1]\}$ in the first case, and $\{x[1], z[1]\}$ in the second.

MH3: Finally, when algorithm *MH3* (see Fig. 5) is called, the graph G only contains variables involved in zero or one constraint, i.e. every variable will be of one of the forms found in Fig. 6. The weights shown in the figure is now added to the corresponding assignments in Θ, the original problem, and the resulting weighted 2-SAT problem is given to a *2-SAT$_w$* solver. If the solution β returned by the solver has weight W, this means that we can add assignments (i.e. vertices) to α and create a solution which differs from β on W assignments in the following way: First of all, since all assignments in α are given weight 0, if any of these are chosen, they will not add anything to the distance, while the other possible value for all these variables will add one to the distance (and are consequently given a weight of 1 on line 3.) For the unconstrained variables in G, i.e. all vertices x with $\delta(x) = (1,1)$, we can choose freely which value they should assume, and thus we can always find an assignment which adds one to the distance from β by choosing the other value for α. The remaining components then consist of pairs of variables with one edge between them, i.e. hurdles. If β contains both assignments with weight $1/2$, then obviously, we have to add one of them to α, since not both assignments with weight $3/2$ are allowed simultaneously – and thus we get a distance of 1, which is the sum of the weights in β. On the other hand, if β contains one $3/2$ and one $1/2$ assignments, then we can choose the opposing value for both of these and get a distance of 2. Consequently, the pair returned on line 12 will have a Hamming distance equal to the weight of β, and with α and G given, no pair with greater Hamming distance can exist.

Except for the call to *2-SAT$_w$* on line 6, every step of algorithm *MH3* can be carried out in polynomial time, thus the time complexity is fully determined by that of the *2-SAT$_w$* algorithm.

Example 3 (cont'd). Assuming we reach algorithm *MH3* with $\alpha = \{x[0], y[1]\}$, i.e. from branch 1 in Example 2 earlier, the algorithm will now assign weights to the assignments in the original problem. Each assignment in α is given weight 0, while its negation is given weight 1. In our case, we get $\boldsymbol{w}(x[0]) = \boldsymbol{w}(y[1]) = 0$, and $\boldsymbol{w}(x[1]) = \boldsymbol{w}(y[0]) = 1$. We have no "hurdles" in our graph (see Fig. 6b), but we have one free variable, z, which we assign weights $\boldsymbol{w}(z[0]) = \boldsymbol{w}(z[1]) = 1$.

Next, we call the weighted 2-SAT solver with the original instance together with the weight vector. It is easy to see that in our case, the maximum weight of a model will be 3; $\beta = \{x[1], y[0], z[1]\}$ has this weight, for instance. Consequently, since z is unconstrained in our graph, we can simply add $z[0]$ to α and we have two satisfying assignment α, β with a hamming distance of 3.

Theorem 1. *Algorithm MH1 correctly solves* MAX HAMMING DISTANCE $(2,2)$-*CSP and has a running time of* $\mathcal{O}\left((a \cdot 1.3803)^n\right)$, *where n is the number of variables in the problem, and $\mathcal{O}(a^n)$ is the time complexity of solving a weighted 2-SAT problem.*

Proof. The highest work factor in algorithms *MH1, MH2, MH3* is $\tau(4,1)$, giving a running time of $\mathcal{O}(1.3803^n)$. The call to the weighted 2-SAT algorithm is done for every leaf in the search tree, and thus we get a total time complexity of $\mathcal{O}\left((a \cdot 1.3803)\right)$ if we assume we can solve weighted 2-SAT in $\mathcal{O}(a^n)$.

In every case, the algorithm branches on both values for a variable, thus from the correctness of the *2-SAT$_w$* algorithm we know that the two solutions returned will be at a maximum hamming distance from each other.   □

Corollary 1. MAX HAMMING DISTANCE $(2,2)$-*CSP can be solved in* $\mathcal{O}(1.7338^n)$.

Proof. Dahllöf et al. [4] presents an algorithm for solving weighted 2-SAT in $\mathcal{O}(1.2561^n)$, and this together with Theorem 1 gives the result.   □

4   Algorithm for Max Hamming Distance (d, l)-CSP

For problems where the arity of the relations is greater than 2, microstructures are not as convenient and we have to find a different approach.

Let us first consider the following problem: Given a CSP instance $\Theta = (X, D, C)$, can we find a pair of solutions with Hamming distance equal to k? One obvious way of doing this is the following:

1. Pick a subset Y of X with $|Y| = k$
2. Create a copy $\Theta' = (X', D, C')$ of Θ over variables X'
3. For each $x \in Y$, add the constraint $x \neq x'$ to C', and
4. for each $x \notin Y$, add the constraint $x = x'$ to C'.
5. If Θ' is satisfiable with solution s
   - Solve the instance $(X \cup X', D, C')$, giving a satisfying assignment s.
   - For each $x \in X$, add $s(x)$ to α
   - For each $x' \in X'$, add $s(x')$ to β
   - Return (α, β)

There are 2^n ways to choose Y on the first line, so if we can solve the satisfiability problem for Θ in time $\mathcal{O}(h(n))$, then, since the number of variables in Θ' is twice that of Θ, we can find a pair of solutions with maximum Hamming distance in $\mathcal{O}(2^n h(2n))$. For example, since 2-SAT can be solved in linear time, we would, using this approach, get a running time of $\mathcal{O}(2^n)$ for the MAX HAMMING DISTANCE $(2,2)$-CSP. This does give a slower running time than the algorithm we presented in the previous section, but it can be applied to CSP instances with domain size and constraint arity greater than 2.

algorithm MAX HAMMING DISTANCE (d, l)-CSP $(\Theta = (X, D, C))$

1. **for** $k := |X|$ **down to** 0 **do**
2.   **for each** $\chi \subseteq X$, $|\chi| = k$ **do**
3.   Let $\Theta' = (X', D, C')$ be a copy of Θ
4.   Let $\gamma \subseteq C$ be all constraints involving variables from χ
5.   Create γ' by exchanging all variables *not* in χ with
            their counterparts from X'
6. $C' := C' \cup \gamma'$
7.   **for each** $x \in \chi$ **do**
8. $C' := C' \cup \{x \neq x'\}$
9.   **if** $(X \cup X', D, C')$ is satisfiable **then**
10.   Let α, β be the two assignments found in a solution to Θ'
11.   **return** (α, β)
12.   **end if**
13.   **end for**
14. **end for**

Fig. 9. Algorithm for MAX HAMMING DISTANCE (d, l)-CSP.

Example 4. Again, consider the instance $(x \vee y) \wedge (\neg x \vee z)$. Following the algorithm in Fig. 9, we begin with trying to determine if there are two solutions with a distance of 3. We get a new, larger instance, which looks as follows:

$$(x \vee y) \wedge (\neg x \vee z) \wedge (x' \vee y') \wedge (\neg x' \vee z') \wedge$$
$$(x \neq x') \wedge (y \neq y') \wedge (z \neq z')$$

This instance has solution $\{x \mapsto 1, y \mapsto 0, z \mapsto 1, x' \mapsto 0, y' \mapsto 1, z' \mapsto 0\}$ and consequently, there are two solutions with a hamming distance of 3 – we get one from reading the values of x, y, z and the other from the values of x', y', z'. Had this instance been unsatisfiable, we would have had to move on to try hamming distance 2, etc.

Actually, it is unnecessary to make a copy of *all* the variables. Having selected k variables that should be different in the two solutions, we only need to make copies of those, leaving the remaining $n - k$ variables unchanged. Thus, we get the algorithm for MAX HAMMING DISTANCE (d, l)-CSP given in Fig. 9.

Theorem 2. *If we can solve (d, l)-CSP in $\mathcal{O}(a^n)$, then there exists an algorithm for MAX HAMMING DISTANCE (d, l)-CSP which runs in $\mathcal{O}((a(1 + a))^n)$.*

Proof. In the algorithm presented in Fig. 9, the instance Θ' will contain $2n - k$ variables, and there are $\binom{n}{k}$ ways of choosing χ. Consequently, given that we can solve (d, l)-CSP in $\mathcal{O}(a^n)$, the algorithm has a total running time of

$$\mathcal{O}\left(\sum_{k=0}^{n} \binom{n}{k} a^{2n-k}\right) = \mathcal{O}\left(a^n \sum_{k=0}^{n} \binom{n}{k} a^{n-k}\right) = \mathcal{O}(a^n (1 + a)^n)$$

and the result follows.   □

In Example 4 we saw how the algorithm for MAX HAMMING DISTANCE (d, l)-CSP worked. On instances of l-SAT we can actually do better than this. Since

algorithm MAX HAMMING DISTANCE $(2, l)$-CSP (F)

1. **for** $k := |\text{Var}(F)|$ **down to** 0 **do**
2.   **for each** $\chi \subseteq \text{Var}(F)$ with $|\chi| = k$ **do**
3.   Let γ be all the clauses of F containing variables from χ
4.   Create $\boldsymbol{\gamma}$ by negating all occurrences of a variable $x \in \chi$ in γ
5.   Let $F^+ := F \cup \{\boldsymbol{\gamma}\}$
6.   **if** F^+ is satisfiable **then**
7.   Let α, β be the two found in a solution to F^+.
8.   **return** (α, β)
9.   **end if**
10.   **end for**
11. **end for**

Fig. 10. The MAX HAMMING DISTANCE $(2, l)$-CSP algorithm.

there are only two possible domain values, and we force x' to always assume the opposite of x, there is no reason to create new variables. Instead, we duplicate the clauses containing variables on which the two solutions should differ, and among these clauses, we replace every literal containing one of these variables with its negation. In the example, we would get:

$$(x \vee y) \wedge (\neg x \vee z) \wedge (\neg x \vee \neg y) \wedge (x \vee \neg z)$$

This formula has a solution $\{x \mapsto 0, y \mapsto 1, z \mapsto 0\}$, and we can easily derive a solution to the original formula which differ on the assignment of all three variables.

As can be seen in Fig. 10, the algorithm for MAX HAMMING DISTANCE $(2, l)$-CSP is similar to the one for the general case, but it does not add any variables to the problem.

Theorem 3. *If we can solve $(2, l)$-CSP in $\mathcal{O}(a^n)$, then there exists an algorithm for solving* MAX HAMMING DISTANCE $(2, l)$-CSP *which runs in $\mathcal{O}((2a)^n)$.*

Proof. The algorithm in Fig. 10 considers all subsets of variables of the problem, as discussed in this section. Consequently, it will deliver a solution in $\mathcal{O}((2a)^n)$ time. $\square$

There exist a number of algorithms for special cases of (d, l)-CSPs, and we can use them in conjunction with Theorems 2 and 3 to get the following corollary:

Corollary 2. *There exist algorithms for solving* MAX HAMMING DISTANCE (d, l)-CSP *with running times*

1. $\mathcal{O}(3.2264^n)$ for $d = 3, l = 2$,
2. $\mathcal{O}((d(0.4518 + 0.2042d))^n)$ for $4 \leq d \leq 10, l = 2$,
3. $\mathcal{O}((d!^{1/d}(1 + d!^{1/d}))^n)$ for $d \geq 11, l = 2$,
4. $\mathcal{O}(2.6604^n)$ for $d = 2, l = 3$,
5. $\mathcal{O}((4 - 4/l + \epsilon)^n)$, for $d = 2, l \geq 4$,
6. $\mathcal{O}(((d - d/l)^2 + d - d/l + \epsilon)^n)$ for $d \geq 3, l \geq 5$.

where $\epsilon > 0$ is an arbitrarily small constant.

Proof. Combine either of Theorems 2 and 3 with

1. the $(3,2)$-CSP algorithm by Eppstein [5],
2. the $(d,2)$-CSP algorithm by Eppstein [5],
3. the $(d,2)$-CSP algorithm by Feder & Motwani [6],
4. the 3-SAT algorithm by Hofmeister *et al.* [8],
5. the l-SAT algorithm by Schöning [11],
6. the (d,l)-CSP algorithm by Schöning [11].

and the result follows.

Acknowledgments

The authors would like to thank Peter Jonsson for useful comments during the writing of this paper.

References

1. O. Angelsmark and J. Thapper. Microstructure based algorithms for three constraint satisfaction optimisation problems, 2004. Unpublished manuscript. Available for download at http://www.ida.liu.se/~olaan/papers/three_algorithms.ps.
2. B. Aspvall, M. F. Plass, and R. E. Tarjan. A linear time algorithm for testing the truth of certain quantified Boolean formulas. *Information Processing Letters*, 8(3):121–123, Mar. 1979.
3. P. Crescenzi and G. Rossi. On the Hamming distance of constraint satisfaction problems. *Theoretical Computer Science*, 288(1):85–100, October 2002.
4. V. Dahllöf, P. Jonsson, and M. Wahlström. On counting models for 2SAT and 3SAT formulae, 2003. Unpublished manuscript. Available for download at http://www.ida.liu.se/~magwa/research/merge23sat.ps.
5. D. Eppstein. Improved algorithms for 3-coloring, 3-edge-coloring, and constraint satisfaction. In *Proceedings of the 12th Annual Symposium on Discrete Algorithms (SODA-2001)*, pages 329–337, 2001.
6. T. Feder and R. Motwani. Worst-case time bounds for coloring and satisfiability problems. *Journal of Algorithms*, 45(2):192–201, Nov. 2002.
7. M. R. Garey and D. S. Johnson. *Computers and Intractability: A Guide to the Theory of NP-Completeness*. W.H. Freeman and Company, New York, 1979.
8. T. Hofmeister, U. Schöning, R. Schuler, and O. Watanabe. A probabilistic 3-SAT algorithm further improved. In H. Alt and A. Ferriera, editors, *Proceedings of the 19th International Symposium on Theoretical Aspects of Computer Science (STACS-2002)*, pages 192–202, Antibes Juan-les-Pins, France, 2002. Springer-Verlag, Berlin, Heidelberg.
9. P. Jégou. Decomposition of domains based on the micro-structure of finite constraint-satisfaction problems. In *Proceedings of the 11th (US) National Conference on Artificial Intelligence (AAAI-93)*, pages 731–736, Washington DC, USA, July 1993. AAAI.
10. O. Kullman. New methods for 3-SAT decision and worst-case analysis. *Theoretical Computer Science*, 223(1–2):1–72, 1999.
11. U. Schöning. A probabilistic algorithm for k-SAT and constraint satisfaction problems. In *40th Annual Symposium on Foundations of Computer Science (FOCS-1999)*, pages 410–414. IEEE Computer Society, 1999.

A System Prototype for Solving Multi-granularity Temporal CSP⋆

Claudio Bettini, Sergio Mascetti, and Vincenzo Pupillo

DICo – Università di Milano, Italy

Abstract. Time granularity constraint reasoning is likely to have a relevant role in emerging applications like GIS, time management in the Web and Personal Information Management applications for mobile systems. This paper reports recent advances in the development of a system for solving temporal constraint satisfaction problems where distance constraints are specified in terms of arbitrary time granularities.

1 Introduction

When variables in a constraint satisfaction problem (CSP) are used to represent event occurrences and constraints to represent their temporal relations, a CSP is called *temporal* CSP or TCSP. Scheduling, planning, diagnosis, natural language understanding, and even temporal databases are examples of areas where temporal CSP's have been applied.

In some cases, a temporal CSP can be formulated in terms of qualitative temporal relations between events, like "event1 must occur *before* event2" or "event1 must occur *immediately after* event2", while in other cases quantitative temporal relations are necessary, like "event2 must occur *at least 1 time unit* and *at most 5 time units after* event1". The many formalisms and algorithms proposed in the literature for TCSP have essentially ignored the subtleties involved in the presence of multiple time units (granularities) in the temporal constraints. Examples of simple constraints specified in terms of a time granularity are the following: "package shipment must occur *the next business day* after check clearance" and "package delivery should occur *during working hours*". There are several emerging applications like GIS, time management in the Web, and Personal Information Management for mobile systems that could greatly benefit of algorithms and systems for time granularity constraint reasoning.

In the last years we have been investigating the concept of time granularity and multi-granularity TCSP, focusing on GSTP, the extension of STP [7] to multiple and arbitrary granularities. Technically, in a GSTP, binary constraints have the form $Y - X \in [m, n]\, G$, where m and n are the minimum and maximum values of the distance between X and Y in terms of time granularity G. Variables take values in the positive integers, and unary constraints can be applied on their domains.

⋆ This work has been partially supported by Italian MIUR (FIRB "Web-Minds" project N. RBNE01WEJT_005).

B. Faltings et al. (Eds.): CSCLP 2004, LNAI 3419, pp. 142–156, 2005.

A first issue in the representation and processing of these constraints is the need for a clear semantics for time granularities. *Business days*, for example, may really have different meanings in different countries or even in different companies. In this respect GSTP adopts a formalism, first introduced in [3], which can model arbitrary user-defined time granularities and has a clear set-theoretic semantics. In order to guarantee a finite representation, granularities in GSTP are limited to those that can be defined in terms of periodic sets. Hours, days, weeks, business days, business weeks, fiscal years, and academic semesters are common examples.

A second issue is related to the difficulty to reduce a network of constraints given in terms of different granularities into an equivalent one with all constraints in terms of the same granularity, so that some of the standard algorithms for CSP could be successfully applied. Indeed, any conversion necessarily introduces an approximation; For example, a constraint imposing delivery to start the next business day may be translated in terms of hours with a minimum of 1 hour and a maximum of 95 hours[1]. However, if the check is cleared on Monday, the constraint in hours would allow a shipment on Thursday which is clearly a violation of the original constraint. Approximate conversion algorithms are extensively discussed in [3,4]. We have shown that any consistency algorithm adopting these conversions as the only tool to reduce the problem to a standard CSP is inevitably incomplete, and have proposed a different algorithm, called AC-G, which has been proved to be complete [5].

A prototype system, named GSTP, has been developed at the University of Milan with the objective of providing universal access to the implementation of a set of algorithms for multi-granularity temporal constraint satisfaction. GSTP, in addition to implementing the reasoning algorithms, assists the user in the definition of constraint networks, in their submission to a remote processing service and in the analysis of the output. A rich pre-defined set of time granularities is available, but new user-defined granularities can be added, and they will be handled by the constraint solving algorithms. The GSTP system has been publicly shown at the Intelligent Systems Demonstration venue at IJCAI 2003 [2]. While the prototype is mainly based on algorithms presented in [5], several enhancements, both theoretical and at the implementation level have been studied and applied. This paper illustrates the overall architecture and essential features of the system, reports recent enhancements and preliminary experimental results.

An extensive literature exists on CSP and TCSP problems. The most popular techniques to deal with CSP are arc- and path-consistency. Several versions of these algorithms have been proposed [8,1]. These algorithms are not specific for temporal constraints and usually assume a finite domain; extensions to deal with infinite domains and TCSP have also been studied [7], but they do not deal with periodic sets. Recent work has been done on identifying tractable classes of periodic CSP [6], but we do not see any immediate applicability of those results to the problems addressed by GSTP. We are also not aware of any CSP

---

[1] The number 95 takes into account a check clearance at the beginning of a Friday and a shipment at the end of next Monday according to the constraint.

system that can directly handle a GSTP problem. Related ideas can be found in [9], where authors note that AC-3 can be generalized to deal with intensionally described domains and constraints, provided that the *domain restriction operation*, usually contained in the REVISE procedure of AC-3, can be performed on the intensional descriptions. Indeed, the AC-G algorithm proposed in our paper can be considered an extension of the AC-3 algorithm to deal with possibly infinite (but periodic) domains and with binary temporal constraints in terms of multiple periodic granularities.

The rest of the paper is organized as follows: In the next section we briefly present the architecture of the GSTP system, and describe two of the three components, the Web Service and a client graphical interface. In Section 3 we present the third component, the constraint solver, illustrating the main algorithm and some recent enhancements. In Section 4 we briefly discuss implementation issues and experimental results, and in Section 5 we conclude the paper.

2   The GSTP Architecture

Figure 1 shows the general architecture of the GSTP system. There are three main modules: the constraint solver, the web service, which enables external access to the solver, and a web service client user interface that can be used locally or remotely to design and analyze constraint networks. All data, including time granularity definitions, constraint network specification, algorithm parameters and processing requests are encoded in XML following specific XML schemas.

The GSTP Constraint Solver is clearly the most complex and innovative component, and the one which required the most implementation efforts; It is described separately in Section 3. Here we give a brief introduction to the main functionalities of the Web Service and of the Client Interface.

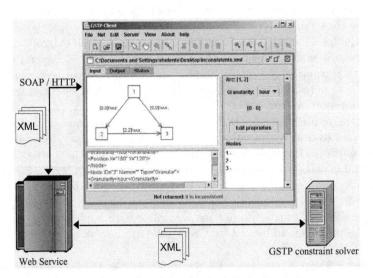

Fig. 1. The GSTP Architecture.

2.1   The GSTP Web Service

The Web Service defines, through a WSDL specification, the parameters that can be passed to the constraint solver, including the XML schema for the constraint network specification. The service is exposed to the public web, and despite we provide a specific client application, it can be invoked by different clients or web applications. Therefore, in principle, our service can be easily integrated in any third party software which requires GSTP processing.

The web service application performs three tasks: first of all it validates the parameters by checking if the XML is valid with respect to the XML schema and if the names of the granularities used are already defined. Then it invokes the solver and finally it passes back the results in XML format.

A single service is currently supported even if a number of parameters can be used to specify different versions of the constraint solver algorithms. It is possible, for example, to give up completeness by selecting a variant of the main algorithm in order to have much lower response time, or to use the complete version and possibly set time-out values different from the default ones.

2.2   The GSTP Client

The main goal of the client interface, in addition to remotely interact with the constraint solver through the web service, is to facilitate two tasks: i) the specification and editing of input networks, and ii) the analysis of processed networks.

For the former task, the GSTP Client supports the user by providing standard functionalities like adding, editing and removing nodes or edges. Networks can also be saved and browsed in XML format.

For the latter, more specialized tools have been developed. In fact the result of the GSTP Constraint Solver is a fully connected network having each arc possibly labeled by one constraint for each of the granularities appearing in the input network. It is clear that is practically infeasible to graphically show all this information in a single screen-shot in a way that is still useful to the user. Therefore some functionalities have been introduced: first of all zooming and scrolling features allow to examine large networks, while nodes can be automatically disposed in order not to overlap with each other or to preserve the position they had in the input network. Moreover it is possible to selectively hide and show information from the network: in particular it is possible to have views of the network in terms of specific single or set of time granularities.

Figure 2 shows the GSTP Client interface showing the result of a GSTP constraint solver computation. The nodes are disposed as they were in the input network, and the only edges that are shown are the ones that were explicit in the input network. All the constraints are hidden except those in terms of the granularity bhday (the business hour day, i.e., the working hours during the business days).

A specific functionality has been introduced in order to show a network solution if the network is found to be consistent (see Figure 3).

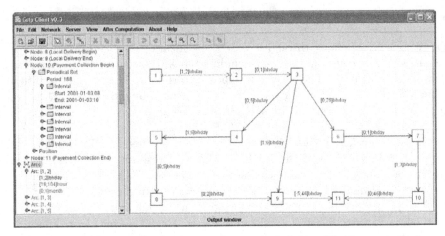

Fig. 2. The view of a processed network in terms of a specific time granularity.

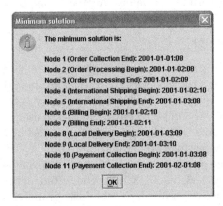

Fig. 3. The minimum solution found by the solver is displayed by the GSTP client interface.

3   The Constraint Solver

The algorithmic task of the constraint solver is to decide the consistency of a set of granularity constraints, to find a solution if one exists, and to restrict the constraints as much as possible while preserving the same set of solutions. Standard algorithms to solve TCSP cannot be easily adapted to GSTP. In order to understand the exact semantics of these networks we first report a few basic notions and then describe our strategy for constraint solving.

3.1   Basic Notions

For lack of space we define informally time granularities referring the interested reader to [3] for formal definitions. A granularity is intuitively seen as a particular grouping of instants from a time domain. Each group is called a *granule*. In most application domains we can find a *bottom* granularity, with the property of being sufficiently fine grained so that no further refinement is needed to represent data in the application domain, and that all other granularities can be represented as groupings of granules of this one. For example, if day is the bottom granularity, week can be defined by grouping 7 days. More complex groupings are needed to represent month, academic-semester, or business-week, but they can all be represented by a periodic expression whose primitive elements are the granules of the bottom granularity. For practical reasons, we index granules with positive integers, so that periodic expressions are in terms of positive integers and algebraic operations can be quite easily defined on granularities. As we will seen in the following, the domains of variables in our constraint networks are indeed indexes of granules of the bottom granularity, and hence, a network solution is the assignment of a specific granule to each variable.

In the following, the function $\lceil t_1 \rceil^G$ denotes the index of the granule of granularity G containing the granule of the bottom granularity indexed by t_1. For example, $\lceil 32 \rceil^{\text{month}} = 2$, i.e., the day indexed by 32 falls in the month indexed by 2, if we assume that the bottom granularity is day, the first day is January 1st and the 1st month is January. Similarly, the function $\lfloor j \rfloor^G$ denotes the set of indexes of granules of the bottom granularity forming the j-th granule of G.

Definition 1. *Let $m, n \in \mathbb{Z} \cup \{-\infty, +\infty\}$ with $m \leq n$ and G a granularity. Then $[m, n] G$, called a* temporal constraint with granularity *(TCG), is the binary relation on positive integers defined as follows: For positive integers t_1 and t_2, $(t_1, t_2) \models [m, n] G$ ((t_1, t_2) satisfies $[m, n] G$) if and only if (1) $\lceil t_1 \rceil^G$ and $\lceil t_2 \rceil^G$ are both defined, and (2) $m \leq (\lceil t_2 \rceil^G - \lceil t_1 \rceil^G) \leq n$.*

Intuitively, for instants t_1 and t_2 of granules of the bottom granularity, if t_1' and t_2' are the indexes of the granules of G containing t_1 and t_2, respectively (i.e., $t_1' = \lceil t_1 \rceil^G$ and $t_2' = \lceil t_2 \rceil^G$), t_1 and t_2 satisfy $[m, n] G$ if the difference of t_2' and t_1' is between m and n (inclusively). That is, the instants t_1 and t_2 are first translated in terms of G, and then the difference is taken. If the difference is at least m and at most n, then the pair of instants is said to satisfy the constraint. For example, if day is the bottom granularity and t_1 and t_2 denote two specific days, the pair (t_1, t_2) satisfies $[0, 0]$ week if the days denoted by t_1 and t_2 are within the same week. Similarly, (t_1, t_2) satisfies $[-1, 1]$ month if those two days are at most one month apart (and the order of them is immaterial). Finally, (t_1, t_2) satisfies $[1, 1]$ year if the day denoted by t_2 is in the next year with respect to the one denoted by t_1.

Definition 2. *A constraint network (with granularities) is a directed graph (W, A, Γ, Dom), where W is a finite set of variables, $A \subseteq W \times W$ a set of arcs, Γ is a mapping from A to the finite sets of temporal constraints with granularities, and Dom is a mapping from W to possibly bounded periodical subsets of the positive integers (indexes of the bottom granularity).*

Note that in these networks multiple constraints (in terms of different time granularities) can be associated with the same arc. From the results in [3], where temporal constraints with granularities were first defined, it follows that, even if constraints on the domains are excluded, and a single TCG is associated with each arc, the consistency problem is NP-hard when arbitrary periodic granularities are allowed, while the single-granularity problem is in PTIME [7].

3.2   A Strategy to Solve GSTP

Constraint satisfaction in GSTP is based on the implementation and optimization of algorithms presented in recent papers. An extension to standard path-consistency based on approximate conversions among constraints with granularities has been proposed in [3]. However, the algorithm was shown to be incomplete with respect to consistency. A sound and complete consistency algorithm, called AC-G (Arc-Consistency with Granularities), has been recently proposed in [5]. On the other side, this algorithm does not directly help in the refinement of constraints and it can greatly benefit from using the previous algorithm as a preprocessing step.

Hence, the solution adopted for the GSTP constraint solver is shown in Fig. 4. In step 1, the original network is decomposed in as many networks as are the granularities appearing in the constraints; each network has the explicit constraints given in terms of one granularity as well as constraints in the same granularity obtained by conversion from others on the same arc, but in terms of different granularities. Then, standard path consistency is applied to each network; networks are re-merged in a single one and if any refinement occurred a new conversion step followed by path consistency is performed, until a fix-point is reached. The resulting network most likely has refined constraints with respect to the original one. Any inconsistency captured by this processing has the effect of terminating the constraint solver reporting the inconsistency status. However, if this is not the case, the network may still be inconsistent and it will go through AC-G (step 2) which is guaranteed to detect an inconsistency if one exists and was not detected by the previous step. From the node domains returned by AC-G, it is possible to further refine some of the constraints (the function doing this job in step 3 is called *RefineArcsFromNodes()*). The steps are repeated, since path consistency applied to the refined constraints may lead to some changes. Correctness and termination have been proved.

Conversion is not a trivial task. A simple example has been given in the introduction considering the conversion of the constraint $[1, 1]$b-day in terms of hours. Since, in general, it is not possible to replace a TCG with an equivalent one in terms of a different granularity, our goal is to find a TCG which is the

---

> **Repeat**
> > 1. **Repeat** Conversion+PC **Until** no change is observed
> > 2. AC-G
> > 3. RefineArcsFromNodes()
> **Until** no change is observed
> **Return** Inconsistent **or** NewNetwork+solution

Fig. 4. The main loop of the constraint solver.

INPUT: a network $\mathcal{N} = \langle W, A, \Gamma, Dom \rangle$.

OUTPUT: a network $\mathcal{N}' = \langle W, A, \Gamma, Dom' \rangle$ equivalent to $\mathcal{N}$ and having one of the domains empty if inconsistent.

METHOD:

$Q := \{(X_i, X_j) \mid (X_i, X_j) \in A\}$

while $Q \neq \emptyset$ **do**

  1. select and delete an arc (X_l, X_k) from Q

  2. **if** $Dom(X_l) \neq^{MAX} Dom(X_l) \cap (Dom(X_k) \uplus \Gamma(X_k, X_l))$ **then**

    2.1. $Q := Q \cup \{(X_i, X_l) \mid (X_i, X_l) \in A, i \neq k\}$

    2.2. $Dom(X_l) := Dom(X_l) \cap (Dom(X_k) \uplus \Gamma(X_k, X_l))$

  3. **if** $Dom(X_l) =^{MAX} \emptyset$ **then** $Q := \emptyset$; $Dom(X_l) := \emptyset$

end while

Fig. 5. The AC-G algorithm.

tightest among those in terms of the new granularity that are implied by the network for the same arc[2]. With respect to previously published algorithms for granularity constraint conversion [3], in the current implementation of GSTP, we use a new algorithm described in [4] that has been proved to derive the tightest converted constraints, and indeed leads to a more effective preprocessing for the constraint solver.

3.3   The AC-G Algorithm

The most challenging part of the system is perhaps the implementation of is algorithm, called AC-G. It is based on arc-consistency, and it is essentially an extension of the AC-3 algorithm [8] to deal with possibly infinite (but periodic) domains and with constraints in terms of multiple periodic granularities. This extension is not trivial since it involves the algebraic manipulation of the mathematical characterization of granularities. AC-G also derives the *minimal* solution for the constraint network.

AC-G is in general exponential in the number of granularities involved in the network, but can be considered to take polynomial time when the time granularities in the constraints are known by the system on which the algorithm is run (i.e., the description of granularities is not given as part of the CSP). Note that most practical applications can satisfy this condition.

A sketch of the AC-G algorithm is reported in Figure 5. Without loss of generality, we assume that for each TCG $[m, n]\, G$ on arc (X_l, X_k), the TCG $[-n, -m]\, G$ exists on arc (X_k, X_l). Basically, the algorithm non-deterministically selects and deletes an arc (X_l, X_k) from a queue (Q) that initially consists of all the arcs, and uses the domain for X_k and the constraints between X_k and X_l to restrict the domain of X_l. If it is restricted, the queue is updated so that any arcs that could lead to further restrictions are re-inserted. Eventually, a fixpoint will be reached and the queue will become empty. Except for the presence

---

[2] By logically implied we intuitively mean that when the TCG is associated with the same arc as the source TCG, the solutions for the original network are still solutions.

of granularity constraints, this is a classical arc-consistency algorithm, in the version known as AC-3 [8]. The central issue in the algorithm is how domains can be restricted considering the granularity constraints associated with the arcs. This is achieved by the operation $Dom(X_k) \uplus \Gamma(X_k, X_l)$ that is defined as returning the set $\{t_l \mid \exists t_k \in Dom(X_k) \wedge (t_k, t_l) \models \Gamma(X_k, X_l)\}$. This ensures that for each value t_l in the domain of X_l, there is a value t_k in the domain of X_k such that (t_k, t_l) satisfies all the constraints on arc (X_k, X_l). The current domain of X_l is then intersected with the set derived by the $\uplus$ operation, since any other values for X_l cannot be part of a network solution. A second issue which distinguishes this algorithm from known arc-consistency algorithms is that we are dealing with possibly infinite periodical domains. This not only requires the implementation of algebraic operations over these sets, but may also pose questions about the termination of the algorithm. Essentially, termination is guaranteed by the fact that the equality and inequality tests in the algorithms are limited by the finite constant MAX.[3] The proper identification of this constant has been a relevant technical issue, since it greatly affects the complexity of the algorithm and it is also essential to guarantee its completeness.

While the AC-G algorithm is exactly the one presented in [5], in Figure 6 we present a variant of the procedure to compute $\uplus$ that has lead to significant performance improvements in the implementation of the constraint solver[4].

Very briefly, in Step 1 the indexes of the starting node which denote granules not included in granules of the constraint granularity are excluded (they could not contribute to the result). In Step 2 we determine for each interval of indexes of the starting node the corresponding interval in the ending node by shifting the interval extremes of the minimum and maximum distances specified in the constraint. The optimization we introduce with respect to [5] is that as soon as one of these resulting intervals covers a span of granules greater than the period we stop the procedure (except for computing global bounds in Step 4). Indeed, since we know that the result is periodic with period P, an interval greater than P implies that the result is G itself, possibly limited by new bounds. Step 3 is executed only when none of the intervals resulting from Step 2 is larger than P. In this case the period characterization of the resulting set is obtained by dropping some granules of G accordingly to the intervals derived in Step 2.

The correctness proof can be trivially reduced to the one provided in [5].

4   Implementation and Experimental Results

In this section we provide some information about the technologies used to implement the GSTP service and report some experimental results.

---

[3] $S_1 \neq^{MAX} S_2$ ($S_1 =^{MAX} S_2$, resp.) means that S_1 and S_2 are different (equal, resp.) if only numbers no greater than MAX are considered.

[4] The version of $\uplus$ reported here assumes a single TCG is associated with each arc. The same procedure can be used in the general case by first transforming the input network into an equivalent one that meets this requirement. Alternatively, a slightly different procedure can be used to operate directly on multi-TCG arcs.

INPUT: a finite or periodical set S and a TCG $[m,n]\,G$.

OUTPUT: the finite or periodical set $S \uplus \{[m,n]\,G\}$

METHOD:

Step 1: Replace S with its intersection with G. If empty, return the empty set.

Step 2: (Any bound on S and G is ignored here and in Step 3)

For each interval $[a_i, b_i]$ in the resulting representation of S

Do:

– compute $[Lo_i, Up_i]$ with $Lo_i = min\lfloor\lceil a_i\rceil^G + m\rfloor^G$, and $Up_i = max\lfloor\lceil b_i\rceil^G + n\rfloor^G$.

– **If** $|Up_i - Lo_i| \geq P - 1$ then **Goto** Step 4 (G is the output set, with bounds as computed in Step 4). **Endif.**

– **If** $Lo_{i-1} \leq Lo_i$ and $\lceil Lo_i\rceil^G \leq \lceil Up_{i-1}\rceil^G + 1$ **Do**

– substitute interval $[Lo_{i-1}, Up_{i-1}]$ in the list with $[Lo_{i-1}, Up_i]$.

– If $|Up_i - Lo_{i-1}| \geq P - 1$ then **Goto** Step 4 (G is the output set, with bounds as computed in Step 4). **EndDo**

Otherwise insert $[Lo_i, Up_i]$ in the list. **Endif.**

EndDo

Step 3: The period representation of the output set is derived from the one of G by excluding each granule $G(j)$ such that there is no $K \in \mathbb{Z}$ and no i for which we have $j = j' + K * R$, $\lceil Lo_1\rceil^G \leq j' < \lceil Lo_1 + P\rceil^G$, and $\lceil Lo_i\rceil^G \leq j' \leq \lceil Up_i\rceil^G$ where R is the number of granules of G in each period, Lo_1 is the first value derived in Step 2 and $[Lo_i, Up_i]$ is the i-th pair of bounds in the list computed in Step 2.

Step 4: The global bounds of the output set are:

$Lo = max(min\lfloor\lceil t_{first}\rceil^G + m\rfloor^G, min\lfloor l\rfloor^G)$, and

$Up = min(max\lfloor\lceil t_{last}\rceil^G + n\rfloor^G, max\lfloor u\rfloor^G)$, where t_{first}, t_{last} are the first and last values in the set S from Step 1, and l, u are the indexes of the first and last granule of G.

Fig. 6. The optimized procedure for $\uplus$.

4.1 Technologies and Optimizations

The Web Service has been implemented in Java using the "Apache Axis" development framework, and Tomcat as the application server. SOAP over HTTP is used to transmit and receive requests. The service is currently active on a publicly accessible server. Java is also the choice for the web service client interface, mostly because of platform independence. It has been tested on recent versions of the Linux and Windows operating systems. On the contrary, the constraint solver has been entirely implemented in ANSI C 99 using the GCC compiler and the libxml2 libraries for parsing XML input. The choice was clearly dictated by the efficiency of the resulting code with respect to other choices.

Regarding optimizations, we worked in particular on the efficient use of memory, devising appropriate data structures and a cache mechanism for granularity internal representation. The cache is initialized after parsing the input constraint network. It is implemented through a hash table having as key the granularity name. Another optimization worth mentioning is a particular ordering of the list

Q of arcs to be considered by the AC-G algorithm. The ordering is based on the computation of the least common multiple among the period of the granularity in the constraint and the periods of the periodical sets for the domains of starting and ending nodes. Arcs with lower values are processed first. This strategy has significantly reduced the processing time.

4.2   Experimental Results

The constraint solver code has been totally rewritten from its first implementation and extensive testing has been performed. In the following we first illustrate how we generated our benchmarks, and then report experimental results both regarding performance and observed upper bounds on the number of executions of the main loops of the algorithm.

Generating Benchmarks. A nontrivial problem is the generation of a significant set of consistent constraint networks. Indeed, the intuitive approach of generating a network by connecting nodes with randomly created constraints leads to a very high rate of inconsistent networks. We addressed this problem by building networks iteratively, ensuring the consistency of the networks obtained at each iteration. The generation process starts from a small consistent network and a node is added at each iteration. This node is connected to a randomly chosen node by labeling the arc with a single-granularity randomly generated constraint C. The GSTP solver is then executed to verify the consistency of the resulting network. If it is not consistent, C is replaced with a new randomly chosen constraint, and this process is repeated until a consistent network is found Then, additional arcs are added between the new node and other randomly chosen nodes, using as labels a relaxed version of the constraints obtained by the GSTP solver that was executed for consistency check[5]. Since the constraints returned by GSTP preserve consistency, their relaxation preserve it as well. Note that, if no relaxation was performed, the generated network would contain a large number of constraints which would not be restricted by path consistency, and this is not realistic for real networks.

Several parameters affect the generation illustrated above, the most relevant being the set of possible granularities and the range of values for the minimum and maximum distance for the randomly generated constraints, and the range of values for the number of arcs added at each iteration. In particular, the last parameter determines the density of arcs in the generated network: we say that a network has a density of $d\%$ if each node connects with about $d\%$ of the other nodes in the network.

The benchmark tests we performed have been obtained generating a great number of networks (tens of thousands). The specific benchmark test considered in this paper includes networks with up to 50 nodes, a density of arcs of 5%, and the granularity of each constraint randomly selected among a set of 10

---

[5] Note that path-consistency is a step of the algorithm, and returns a set of constraints (each one in a different granularity) for each pair of nodes in the network. We currently take one randomly chosen constraint from each set.

different granularities. Granularities in this set include standard ones with their exceptions (e.g., years modeling leap years), granularities with non-contiguous granules (like business days), and granularities whose granules have internal gaps (e.g., business months).

A different range of values for minimum and maximum distances for each randomly generated constraint is specified depending on the granularity of the constraint. Ranges are chosen trying to simulate realistic problem specifications, and taking into account that small distance values in terms of granularities with large granules are converted during constraint processing into large distance values in terms of granularities with small granules. For example, a maximum distance of $+100$ in terms of years becomes $+885383$ in terms of hours. In order to understand the impact of this parameter, we performed a specific test with two sets of networks with the second one generated using ranges which are larger by a order of magnitude with respect to the ones used for the first set.

Performance Results. All the performance experimental results reported here are obtained by averaging the processing time over several networks generated using the same parameters. As expected, the processing time is significantly affected by the number of nodes in the network. However, experimental results have shown that a very relevant parameter is the subset of granularities appearing in the constraints. When the least common multiple of their periods is large, the processing time sharply increases, and the size of the network in terms of number of nodes becomes less relevant.

Figure 7 shows the difference in performance between networks involving a set of granularities having a high common period, with respect to networks with a low common period.

The chart presents the processing time for four classes of networks differing in the granularities used in their constraints. The first class has all constraints in terms of the granularity **hour** and the common period is clearly 1. The second class has constraints in terms of granularities **hour** and **day**, with their common

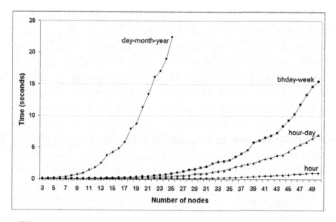

Fig. 7. Computation time for multi-granularity networks.

period being 24. The granularities appearing in the third class of networks are bhday and week, and their common period is 168. Finally, the fourth class of networks involve day, month, and year with a common period[6] of 35064. The chart shows that the processing time grows significantly with the value of the period. In particular, for the fourth class of networks, the processing time with 25 nodes is greater than the processing time of a 50 nodes network of the third class.

Figure 8 shows how the constraints distance range affects processing time. The two classes of networks considered in Figure 8 have constraints in terms of the same set of granularities (bhday and week), but the second class has been generated using distance ranges that are larger than the ones used for the first class by one order of magnitude. It is clearly visible how this parameter affects processing time. Accordingly to our experiments, the processing time growth is similar for classes of networks with a single granularity and in classes of networks using multiple granularities.

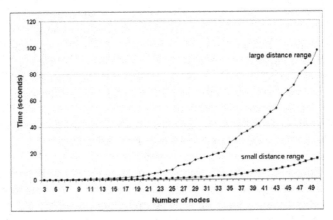

Fig. 8. Impact of the distance range on processing time.

We should also mention that first evaluations on the impact of arc density show very moderate variations of the processing time with changes to this parameter. While we are working on several optimizations that will definitely reduce the execution times measured for this benchmark, the optimizations will not affect the qualitative observations reported above on the main parameters affecting GSTP performance.

Experimental Results on the Upper Bound for Loop Cycles. The extensive testing performed with our benchmarking suite has shown that the number of iterations of Conversion and Path Consistency, i.e., the loop at Step 1 of the algorithm in Figure 4, is very low. Indeed, this number has never been more than 5 in our experiments.

---

[6] We consider leap years, but ignoring exceptions to the 4 years periodicity.

We have also observed that the number of iterations of the external loop of the algorithm, i.e., the three main steps in Figure 4, is very low. Actually, it has never been more than 2 in our experiments. Note that any network inconsistency is always detected at the first iteration, either by Conversion+PC or by the first run of AC-G.

These results are very significant to us since the current theoretical bounds on the number of iterations depend on the range of values in the constraints. They indicate that, despite our efforts in this direction, a much better theoretical characterization of the bounds on iterations can probably be obtained.

5 Conclusions

In this paper we presented the GSTP system, a web service to solve TCSP expressed in terms of multiple time granularities. It is to our knowledge the only implementation of a complete algorithm for consistency checking of these networks. We are currently working in two directions. On one side we are still working at the optimization of reasoning algorithms, while on the other side we are enhancing the web service XML-based architecture to provide easy access to the algorithms by external applications, and to facilitate the specification of user-defined time granularities to be used in GSTP.

Acknowledgments

Many people contributed to the implementation of the GSTP system. In particular we would like to thank Carlo Cestana for his work on the GSTP web service, Simone Ruffini for his work on granularity constraint conversions, and Sean Wang for his valuable suggestions on implementation issues. This work has been partially supported by Italian MIUR (FIRB "Web-Minds" project N.RBNE01WEJT_005).

References

1. C. Bessière. Arc-Consistency and Arc-Consistency Again. *Artificial Intelligence* 65(1):179–190, 1994.
2. C. Bettini, S. Mascetti, V. Pupillo. GSTP: A Temporal Reasoning System Supporting Multi-Granularity Temporal Constraints, in Proc. of Int. Joint Conference on Artificial Intelligence (IJCAI), (Intelligent Systems Demonstrations), pp. 1633-1634, Morgan Kaufmann, 2003.
3. C. Bettini, X. Wang, S. Jajodia. A General Framework for Time Granularity and its Application to Temporal Reasoning. *Annals of Mathematics and Artificial Intelligence* 22(1,2):29–58, 1998.
4. C. Bettini, S. Ruffini, Granularity Conversion of quantitative temporal constraints DICo – Università di Milano, Technical Report N. 276-02, http://homes.dico.unimi.it/~bettini/papers/tr276-02.pdf. Under journal review.

5. C. Bettini, X. Wang, S. Jajodia, Solving Multi-Granularity constraint networks, *Artificial Intelligence*, 140(1-2):107–152, 2002.
6. Hubie Chen, Periodic Constraint Satisfaction Problems: Polynomial-Time Algorithms. In Proc. of Int. Conf. on Principles and Practice of Constraint Programming, LNCS 2833, pp. 199–213, Springer, 2003.
7. R. Dechter, I. Meiri, J. Pearl. Temporal constraint networks. *Artificial Intelligence* 49:61–95, 1991.
8. A. Mackworth, E. Freuder. The complexity of some polynomial network consistency algorithms for constraint satisfaction problems. *Artificial Intelligence* 25:65–74, 1985.
9. A.K. Mackworth, J. A. Mulder, W. S. Havens. Hierarchical Arc Consistency: Exploiting Structured Domains in Constraint Satisfaction Problems. *Computational Intelligence*, 1:118–126, 1985.

Computing Equilibria Using Interval Constraints

Lucas Bordeaux[1,*] and Brice Pajot[2]

[1] DIS, Univ. di Roma "La Sapienza", Italy
bordeaux@dis.uniroma1.it
[2] LINA-CNRS FRE 2729, Univ. de Nantes, France
pajot@lina.univ-nantes.fr

Abstract. Finding NASH equilibria is a hard computational problem which is central to game theory and whose applications range from decision-making to the analysis of multi-agent systems. Despite considerable recent interest and significant recent improvements, the problem remains essentially open in the case of n-person games. We investigate the use of interval-based constraint solving techniques to compute equilibria. We report on experiments made using several encodings of randomly-generated games into continuous CSP, and draw conclusions regarding both the scalability of interval methods for game-theoretic applications and the impact of the symbolic representation of polynomials and of the choice of the propagation technique on the speed of resolution.

1 Introduction and Motivation

Game theory [14] studies situations in which two or more agents with *conflicting interests* are interacting. It provides mathematical arguments to determine which strategy an agent should choose to maximise its profit, and to predict which global scenarios are most likely to arise if all agents behave rationally. As a tool for decision-making, game-theory is complementary to traditional constraint solving and optimisation frameworks in that it can take into account the behaviour of competitors or opponents.

Emerging from research in economics, game theory also found considerable applications to other scientific areas and to computer science in particular. For instance, it is used in the analysis of multi-agent systems [5] and it is a key ingredient in the emerging theory of the Internet [16]. We focus here on the *computational* problems raised by game-theory, of which the most prominent is undoubtedly the computation of NASH equilibria. An equilibrium is a situation in which no agent can increase its benefits by changing his strategy unilaterally. It is therefore natural in many cases to expect that rational behaviour will lead to equilibria. The computation of equilibria is a challenging problem which has recently received interest from the artificial intelligence community (see, *e.g.,* [7, 11, 8, 20, 4, 3, 2, 10, 17]) but, surprisingly, we are not aware of any work on equilibria based on constraint programming.

---

* Work partially supported by project ASTRO funded by the Italian Ministry for Research under the FIRB framework (funds for basic research).

B. Faltings et al. (Eds.): CSCLP 2004, LNAI 3419, pp. 157–171, 2005.

Interval constraint solving [1] is an approach to the resolution of real-valued nonlinear systems based on *interval propagation*. Among other advantages, it shows in some cases superior efficiency to classical mathematical programming methods, it can deal with a large class of constraints (possibly non polynomial ones) and, thanks to a clever use of outward-rounding, it gives correct results (no solution is lost) despite of the finiteness of number representation in floating-point arithmetics. A good example of freely available interval constraint-solving software tool is realpaver[1], developed by Laurent GRANVILLIERS [6].

Although a very promising technology for solving a wide range of important non-linear problems, interval constraint solving has so far not found as many industrial-size applications as discrete constraint solving. Recent successes in such application areas as computer-aided design are starting to change this situation, and new application areas would contribute to the dissemination of this technology. We claim that the application of interval constraint solving to the large avenue of game-theoretic problems is promising, and our purpose here is to initiate its investigation.

Our approach is, more specifically, to focus on the problem of computing NASH equilibria in n-person games using interval constraints. Since the problem has not been addressed yet in the constraint programming literature, we give a self-contained and CP-oriented presentation of the needed game-theoretic material (Section 2), putting emphasis on the encoding of games into CSP (Section 3). Based on specificities of interval-based methods which we introduce in Section 4, we then propose a number of improved variants of this encoding. We describe an implementation based on realpaver and report on our experimentations (Section 5). Our conclusions and perspectives are summarised in Section 6.

2   Games and Equilibria

The simplest model in game theory assumes that each player has a finite number of possible *strategies*, and that the reward of each player can be evaluated by a numerical value (say a real number) called *payoff* which depends on the combination of the strategies chosen by all players.

2.1   An Introductory Example

Basic game-theoretic notions are best understood using examples. Consider for instance a situation in which two agents have to choose between 3 strategies. The strategies of player 1 are called a, b and c and those of player 2 are called α, β and γ. The payoff expected by both players is represented by the following tables (left-hand-side for player 1, right-hand-side for player 2). For instance, if 1 and 2 respectively choose strategies b and γ, their respective payoffs will be 55 and 45. The goal of each player is to maximise its outcome, and decision is taken secretly, so each player has to anticipate what the other can do.

---

[1] http://www.sciences.univ-nantes.fr/info/perso/permanents/granvil

	α	β	γ
a	35	60	15
$1:b$	45	50	55
c	40	70	10

	α	β	γ
a	65	40	85
$2:b$	55	50	45
c	60	30	90

$$(1)$$

Classical interpretations are, for instance, that strategies correspond to propositions in an electoral programme (and payoffs to expected voting scores), to investment policies (and payoff reflect market share), or that they represent military options (and payoffs estimate the probability of victory).

Pure Equilibria. Although no strategy is clearly preferable to all the others in all cases (for neither player 1 nor player 2), it is clear that player 2 should not choose strategy β, because α is a better choice in any case (strategy β is said to be *dominated* by α, domination makes decisions easier but it arises only in very particular cases). Now in the simplified game where strategy β is ignored, player 1 knows what to do, because b is the best choice in both of the remaining cases. But the best player 2 can do then is choose strategy α, and the players should therefore choose the combination $\langle b, \alpha \rangle$. What characterises this rational choice is that each of the strategies b and α is the *best response* to the other: if player 1 chooses strategy b, the best player 2 can do is choose α; if 2 chooses α, the best player 1 can do is choose b. Such a situation is called a (NASH) *equilibrium*, a most important game-theoretic notion acknowledged as a key-concept in the modelling of rational behaviour.

Mixed Equilibria. In our case, the equilibrium was *pure* in the sense that each player could choose one unique strategy, on which all of his effort would concentrate. This corresponds to situations where resources are indivisible. If one can split his investment between several strategies, or makes a choice randomly according to some probability distribution it fixes in advance, the situation gets a bit more complicated. For the sake of readability, consider the simpler game:

	α	β
a	2	0
$1:b$	0	1

	α	β
a	1	0
$2:b$	0	2

$$(2)$$

Two pure equilibria in this game are obviously $\langle a, \alpha \rangle$ and $\langle b, \beta \rangle$. Now, what if each player decides to toss a coin to determine its strategy (*i.e.*, each strategy has probability $\frac{1}{2}$)? Clearly, knowing that player 2 will toss a coin, the payoff for player 1 is defined as:

$$\sigma_1(a). \left(\frac{1}{2}.2 + \frac{1}{2}.0 \right) + \sigma_1(b). \left(\frac{1}{2}.0 + \frac{1}{2}.1 \right) \quad = \quad \sigma_1(a) + \frac{1}{2}.\sigma_1(b)$$

where $\sigma_1(x)$ represents the probabilities given to each strategy $x \in \{a, b\}$ by player 1. Choosing $\sigma_1(a) = \frac{1}{2}$ and $\sigma_1(b) = \frac{1}{2}$ leads to a payoff of $\frac{3}{4}$ for player 1, whereas the strategy $(\sigma_1(a) = 1, \sigma_1(b) = 0)$ would be more appealing (payoff of 1). Therefore we have an unstable situation, where the two strategies are not best responses to each other. On the contrary, if player 1 chooses the distribution $(\sigma_1(a) = \frac{2}{3}, \sigma_1(b) = \frac{1}{3})$ and player 2 chooses $(\sigma_2(\alpha) = \frac{1}{3}, \sigma_2(\beta) = \frac{2}{3})$, the expected payoff of player 1 becomes $\sigma_1(a).(\frac{1}{3}.2 + \frac{2}{3}.0) + \sigma_1(b).(\frac{1}{3}.0 + \frac{2}{3}.1) = \frac{2}{3}(\sigma_1(a) + \sigma_1(b))$ (and symmetrically for player b). Since $\sigma_1(a)$ and $\sigma_1(b)$ sum up to 1, player 1 has no reason to deviate from its strategy which is as maximal as any other strategy would be. Each choice is therefore a best response to the other: by choosing the combination $(\frac{2}{3}, \frac{1}{3})$, each player fixes the payoff of its opponent to $\frac{2}{3}$ in any case. This is but a particular situation leading to an equilibrium on distributions, which is called a *mixed* equilibrium. Note that we have presented examples of games with 2 players, obviously the concept generalises to n-players.

2.2   Definitions of Games and Equilibria

Games. As illustrated in the previous subsection, the games considered in classical computational game theory are composed of *players*, each having a set of (pure) *strategies*, and of a means to evaluate the *payoff* of each combination. The general formulation of games is the following:

Definition 1. [game] *A game is a triple* $\langle \mathcal{N}, \mathcal{S}, \mathcal{P} \rangle$ *where:*

- $\mathcal{N} = \{1, \ldots, n\}$ *is a finite set of* players, *each of which is identified by (and with) its number;*
- $\mathcal{S} = \{S_1, \ldots, S_n\}$ *defines a set* S_i, *called* strategy set, *for each player i. Each* S_i *is a finite set of names called* (pure) strategies;
- $\mathcal{P} = \{p_1, \ldots, p_n\}$ *is a set of* payoff functions *(again, one for each player), each* p_i *is a function of signature* $S_1 \times \ldots \times S_n \to \mathbb{R}$.

Structured Games. An important question is how the payoff functions are defined. In the case of two-person games, the payoff functions can be defined by a pair of matrices (*a.k.a. bimatrix*). In the general case, n-dimensional arrays would be of size $\prod_{i=1..n} |S_i|$, which is exponential in n. It has been noticed by several authors that realistic representations should reflect the locality of the decisions taken by each player: in a system involving a large number of agents, each of them typically takes its decisions according to a restricted subset of the others. In other words, each p_i function is indeed a function of signature $S_{d_1} \times \ldots \times S_{d_p} \to \mathbb{R}$ for a selected subset of indices $\{d_1, \ldots, d_p\} \subseteq \{1, \ldots, n\}$. This type of games is equivalent to the *graphical games* studied in [7,11].

Definition 2. [structured game] *A structured game is a triple* $\langle \mathcal{N}, \mathcal{S}, \mathcal{D}, \mathcal{P} \rangle$ *where* $\mathcal{N} = \{1, \ldots, n\}$ *is a finite set of* players, $\mathcal{S}$ *defines the strategy sets and:*

- $\mathcal{D} = \{D_1, \ldots, D_n\}$ *is the set of* dependencies, *each* $D_i \subseteq \{1, \ldots, n\}$ *represents the subset of players whose decision impacts player i's payoff;*
- $\mathcal{P} = \{p_1, \ldots, p_n\}$ *is a set of* payoff functions, *each* p_i *is a function of signature* $S_{d_1} \times \ldots \times S_{d_p} \to \mathbb{R}$ *where* $D_i = \{d_1, \ldots, d_p\}$.

In all practically relevant cases, we have $i \in D_i$ (the payoff of a player depends, in particular, on its own choices).

Equilibria. We shall call (pure) *situation* a vector $s = \langle s_1, \ldots s_n \rangle$, where $s_i \in S_i$, which gives the pure strategy chosen by every player. Payoff functions therefore assign a (positive real) number to each situation.

Slightly abusing notation for the sake of readability, we note $p_i(s)$ the payoff of player i even though not all the values s_i are actually taken into consideration.

Definition 3. [pure equilibrium] *A (pure) situation $s = \langle s_1, \ldots, s_n \rangle$ is a (pure) equilibrium if the choice s_i made by each player i maximises $p_i(s)$ i.e., $\forall i \in 1..n$:*

$$p_i \langle s_1, \ldots, s_n \rangle = \max\{p_i \langle s_1, \ldots, s_i', \ldots, s_n \rangle \mid s_i' \in S_i\}$$

Defining *mixed* equilibria just complicates the notation a little more, and requires defining mixed strategies based on *probability distributions*. We denote probability distributions over a (typically finite) set A by $Dist(A)$, i.e., $Dist(A)$ denotes the set of mappings f from A into $[0,1]$ s.t. $\sum_{v \in A} f(v) = 1$.

Definition 4. [mixed strategy, mixed situation] *A mixed strategy for player i is a probability distribution over the set of pure strategies of i, i.e., a function $\sigma_i \in Dist(S_i)$. A mixed situation is a vector $\sigma = \langle \sigma_1, \ldots, \sigma_n \rangle$, which gives the mixed strategy chosen by every player.*

The payoff of a mixed situation is determined by the payoff of each pure situation as follows (abusing notation again, we also note mixed payoffs as p_i):

Definition 5. [mixed payoff] *The payoff of a mixed situation $\sigma = \langle \sigma_1, \ldots, \sigma_n \rangle$ for player i is defined by:*

$$p_i(\sigma) = \sum_{s \in S_1 \times \ldots \times S_n} \left(p_i(s).\pi(s) \right)$$

where $\pi(s) = \prod_{j \in 1..n} \sigma_j(s_j)$ denotes the probability of situation s.

Definition 6. [mixed equilibrium] *A mixed situation $\sigma = \langle \sigma_1, \ldots, \sigma_n \rangle$ is a mixed equilibrium if the choice σ_i made by each player i maximises $p_i(\sigma)$ i.e., for each $i \in \mathcal{N}$:*

$$p_i \langle \sigma_1, \ldots, \sigma_n \rangle = \max\{p_i \langle \sigma_1, \ldots, \sigma_i', \ldots, \sigma_n \rangle \mid \sigma_i' \in Dist(S_i)\}$$

In the case of structured games, each player only imposes maximality w.r.t. the set $D_i = \{d_1, \ldots, d_p\}$ of players it depends on, which just complicates a bit the notation (we assume $i = d_a$):

$$p_i \langle \sigma_{d_1}, \ldots, \sigma_{d_p} \rangle = \max\{p_i \langle \sigma_{d_1}, \ldots, \sigma_{d_a}', \ldots, \sigma_{d_p} \rangle \mid \sigma_{d_a}' \in Dist(S_i)\}$$

2.3   State of the Art on Computing Equilibria

Pure equilibria are not guaranteed to exist, and they do not raise interesting computational issues (they can be computed by simply enumerating the pure strategies, *i.e.*, the cells of the tables). On the contrary, mixed equilibria provably exist for every game, and there may even exist many of them, possibly with different payoffs, as proved by NASH [15]. But their computation is difficult and no polynomial-time algorithm is known which solves the problem in general. Two cases are more favourable: 2-person games which are *0-sum* (*i.e.*, in which the sum of the payoffs of the two opponents is constant) can be solved by linear programming and are therefore tractable. For 2-player games in general, an algorithm due to HOWSON and LEMKE works well in practice [21]. Like the simplex, to which it is similar, its worst-case runtime is however provably exponential[2].

The problem of efficiently computing equilibria for n-person games is essentially open, and it is clear that our effort should focus on this problem. Several algorithms have been proposed in the literature [13] but their scalability is so far acknowledged to be insufficient. DATTA [4] reports on her experiments with the prominent game-theoretic software package gambit [12] as follows: *the only games which it was able to solve with any consistency were* [...] *3 players, each with two pure strategies.* Very recent work from the AI community [7, 11, 8, 17] have lead to improvements and shown, in particular, that continuation methods are promising [4, 2]. It is clear that other types of methods can potentially contribute to more robust and scalable algorithms and should therefore be investigated.

We additionally note that equilibria in n-person games can be irrational numbers. It is therefore *not even possible to exactly compute them* in general (a recent result [9] states that every game has at least one equilibrium where all values are *algebraic*, but its representation can nonetheless be exponentially large.). The only safe way we know to circumvent the incorrectness of finite machine representations like floating-point numbers is to use interval arithmetics.

3 Encoding Games as Continuous CSP

In this section we describe a translation of games into continuous CSP. This encoding, which is based on a classical theorem by NASH, was therefore implicit in the game theory literature and is also used in recent theoretical work [9], yet it does not seem to have been considered in recent AI works. Its main advantage is its simplicity.

3.1 Ingredients of the Encoding

We recall that a (continuous) CSP as defined in *e.g.*, [1] is a triple $\langle \mathcal{V}, \mathcal{I}, \mathcal{C} \rangle$, where $\mathcal{V}$ is a set of variables, $\mathcal{I} = \{I_x \mid x \in \mathcal{V}\}$ is a set of domains (each I_x represents the interval of values that variable x can take) and $\mathcal{C}$ is a set of constraints (here polynomial), each of which relates some of the variables of $\mathcal{V}$.

Variables and Constraints on Distributions. We want to find *mixed strategies* which respect some constraints. The variables of the problem should therefore represent the probability distributions:

$$\sigma_1 \in S_1 \to [0,1] \quad \ldots \quad \sigma_n \in S_n \to [0,1]$$

---

[2] An algorithm whose complexity is sub-exponential has recently been proposed for computing approximate equilibria (*a.k.a.* ϵ-equilibria) in 2-person games, thanks to a proof of existence of supports of logarithmic size [10].

Of course, since each S_i is a small finite set, we use an array of variables valued in $[0, 1]$ for each σ_i, *i.e.*, for each i we define a set of variables $\sigma_i[v]$, for $v \in S_i$. A first set of constraints is needed (together with the domain declarations bounding the variables to $[0, 1]$) to state that each σ_i is a probability distribution:

$$\bigwedge_{i \in 1..n} \left(\textstyle\sum_{v \in S_i} \sigma_i[v] = 1 \right)$$

Maximality Constraints. The main ingredient we have to define left is the set of constraints expressed by each player to impose that its strategy be a best response to the others. Informally, these constraints have the following form:

$$\bigwedge_{i \in 1..n} \left(\begin{array}{l} p_i \langle \sigma_1, \ldots, \sigma_n \rangle = \\ \max\{p_i \langle \sigma_1, \ldots, \sigma_i', \ldots, \sigma_n \rangle \mid \sigma_i' \in Dist(S_i)\} \end{array} \right)$$

where each p_i is translated into a polynomial expression over the variables representing the probability distributions of each player. This is clearly an unusual type of constraint, which captures the essence of the computational task at hand. Note that, although the problem has a clear optimisation flavour, it cannot directly be mapped into an optimisation problem (*i.e.*, maximising one function under a set of constraints); what we have instead is a set of *maximality constraints*: each player imposes that its probability distribution maximise its payoff. How these unusual constraints are translated into conventional CSP is what we discuss next.

3.2   Encoding the Maximality Constraints

A simple but completely developed example helps visualising. Let's go back to our game with 2 players and 2 strategies (Sec. 2.1, equation 2). Each player wants to maximise its own payoff function which depends on variables $\sigma_1[a]$, $\sigma_1[b]$, $\sigma_2[\alpha]$, $\sigma_2[\beta]$, but player 1 can choose optimal values for array σ_1, while player 2 sets the values for σ_2. For instance, player 1's constraints have to express the following condition:

$$\left(\begin{array}{l} 2\,\sigma_1[a]\,\sigma_2[\alpha] \\ +\,1\,\sigma_1[b]\,\sigma_2[\beta] \end{array} \right) = \max\left\{ \left(\begin{array}{l} 2\,\sigma_1[a]\,\sigma_2[\alpha] \\ +\,1\,\sigma_1[b]\,\sigma_2[\beta] \end{array} \right) \;\middle|\; \begin{array}{c} \sigma_1[a] \in [0,1], \sigma_1[b] \in [0,1] \wedge \\ \sigma_1[a] + \sigma_1[b] = 1 \end{array} \right\}$$

The meaning of maximality constraints can be expressed using universal quantified constraints similar to those studied, for instance, in the work of Stefan RATSCHAN [18]. More precisely, we need quantifiers on probability distributions, so the first constraint could be expressed as:

$$\forall \sigma_1'[a] \in [0,1]\ \forall \sigma_1'[b] \in [0,1] \quad \left(\begin{array}{l} (\sigma_1'[a] + \sigma_1'[b] = 1)\ \rightarrow \\ \left(\begin{array}{l} 2\sigma_1[a]\sigma_2[\alpha] + 1\sigma_1[b]\sigma_2[\beta] \\ \geq 2\sigma_1'[a]\sigma_2[\alpha] + 1\sigma_1'[b]\sigma_2[\beta] \end{array} \right) \end{array} \right)$$

and symmetrically for the second player. This kind of constraints would be extremely hard to solve in this form, not only because of quantifiers, but also because of the implication connectors which can typically be handled on discrete domains by reification, but are much trickier when the left-hand part is a continuous equality.

Fortunately, a finite number of points suffices to guarantee the universally quantified statement. In our case, it is sufficient to take values $(1,0)$ and $(0,1)$ for $(\sigma'_1[a], \sigma'_1[b])$ because all other values summing to 1 (*e.g.*, 0.5, 0.5) will result in a lower payoff. We obtain the following constraints for player 1:

$$\begin{pmatrix} 2\sigma_1[a]\sigma_2[\alpha] + 1\sigma_1[b]\sigma_2[\beta] \\ \geq 2\sigma_2[\alpha] + 0\sigma_2[\beta] \end{pmatrix} \quad \wedge \quad \begin{pmatrix} 2\sigma_1[a]\sigma_2[\alpha] + 1\sigma_1[b]\sigma_2[\beta] \\ \geq 0\sigma_2[\alpha] + 1\sigma_2[\beta] \end{pmatrix}$$

A general statement of what was exemplified here is the following proposition:

Proposition 1. [alternative characterisation of mixed equilibria] *A mixed situation $\sigma = \langle \sigma_1, \ldots, \sigma_n \rangle$ is a mixed equilibrium iff, for each player $i \in 1..m$, we have:*

$$\bigwedge_{j \in S_i} \left(p_i \langle \sigma_1, \ldots, \sigma_n \rangle \geq p_i \langle \sigma_1, \ldots, \sigma_{i-1}, pure_{ij}, \sigma_{i+1}, \ldots, \sigma_n \rangle \right)$$

where $pure_{ij}$ denotes the distribution which assigns probability 1 to strategy j of player i and 0 to its other strategies.

Proof. (sketched) When the strategies of the other players are fixed, the payoff for i is a weighted average of the payoffs obtained for each of its pure strategies. If the chosen distribution gives a superior payoff to those obtained for every pure strategy, it is therefore also superior to any weighted average of them.

This alternative characterisation was first used by NASH [15] on the way to his proof of the existence of equilibria. The characterisation can indeed be used to construct a continuous function whose fixpoints are the NASH equilibria (such fixpoints provably exist since the distribution space is a convex body). From this characterisation we can derive the encoding formalised in Fig. 1 for unstructured games (in the case of structured games, the notation is just a bit more trickier because we have to define $A = S_{d_1} \times \ldots \times S_{d_p}$.).

3.3   Other Encodings

Several approaches to find equilibria in the recent literature were based on an encoding into another formalism. DATTA [4] uses an encoding which characterises a particular type of equilibria, called *totally mixed*, in which the payoffs for all pure strategies are equal. VICKREY & KOLLER [20] use a discretisation of the problem from which they derive an encoding as a discrete CSP. BLUM *et al.* use a more complex encoding suitable to the continuation methods they consider, and which is based on both a function similar to that of NASH's theorem and an other function called retraction operator. Surprisingly, we are not aware of attempts to

variables $\bigcup_{i \in 1..n} \sigma_i$ where each $\sigma_i = \{\sigma_i[v] \mid v \in S_i\}$

domains $\bigwedge_{i \in 1..n} \left(\bigwedge_{v \in S_i} \sigma_i[v] \in [0,1] \right)$

constraints   *constraints on probability distributions:*
$\bigwedge_{i \in 1..n} \left(\sum_{v \in S_i} \sigma_i[v] = 1 \right)$

maximality constraints:

$$\bigwedge_{i \in 1..n} \bigwedge_{j \in S_i} \left(\begin{array}{l} \displaystyle\sum_{s \in A} p_i(s).\pi(s) \geq \sum_{s \in A^j} p_i(s).\pi^j(s) \\[2ex] \text{with :} \\[1ex] \bullet\ A = S_1 \times \ldots \times S_n \\ \bullet\ \pi(s) = \prod_{k \in 1..n} \sigma_k[s_k] \\[1ex] \bullet\ A^j = S_1 \ldots S_{i-1} \times \{j\} \times S_{i+1} \ldots S_n \\ \bullet\ \pi^j(s) = \prod_{k \in 1..n} coef(k) \\ \text{where } coef(k) := (\text{if } k = j \text{ then } 1 \text{ else } \sigma_k[s_k]) \end{array} \right)$$

Fig. 1. Encoding games into CSP using maximality constraints.

directly use the encoding of maximality constraints to compute NASH equilibria. Its main benefits are that it is the simplest and most natural encoding we have encountered so far and that it is therefore a good basis from which we can derive slightly improved encodings, see next section.

4   Improving the Encoding with Symbolic Transformations

Since we have translated games into continuous CSP, we can now use interval constraint solving algorithms to solve the resulting CSP. The efficiency of these methods is *syntax-dependent*, and rewriting the constraints in an equivalent, but more propagation-friendly form typically speeds up the resolution.

4.1   Interval Propagation Techniques

The paper being intended for an audience with a CP background, we insisted on game-related definitions but will only give a high-level and intuitive description of the algorithms we have used; we refer the reader to *e.g.*, [1,6] for a more detailed presentation of interval constraint programming. Interval CP techniques solve constraints where each variable x_j ranges over an interval I_j (which, for the variables of our encoding, is initially set to $[0,1]$). These intervals define a "box" $B = I_1 \times .. \times I_q$ in which solutions are searched. If $I_j = [l, r]$, the two boxes $I_1..I_{j-1} \times [l,l] \times I_{j+1}..I_q$ and $I_1..I_{j-1} \times [r,r] \times I_{j+1}..I_q$ obtained by fixing x_j to, respectively, its lower bound and its upper bound, are called *facets* of the box B on variable x_j.

Interval CP uses a *branch & prune* approach to find solutions inside the box: the initial box is recursively split into several pieces until boxes which enclose the solutions with satisfactory precision are obtained. To avoid exploring an exponential number of boxes, each constraint of the problem is used to perform a *filtering* of the box, *i.e.*, to bring its facets closer, eliminating regions which do not contain solutions. The two filtering techniques we have experimented are:

Basic interval propagation (*a.k.a. hull consistency*, or *2B-consistency*). It uses simple, multidirectional rules which, for each operator $\diamond \in \{+, -, \star, \ldots\}$, specify how to recompute the interval associated to one of the subterms l, r and up of an expression $l \diamond r = up$ from the two others subterms. For instance, a constraint $x + y = z$ with $I_y = [l_y, r_y]$ and $I_z = [l_z, r_z]$ can be used to suppress from I_x all values not in $[l_z - r_y, r_z - l_y]$.

Box consistency, a technique which achieves a tighter filtering at the price of requiring more computations. It is based on the fact that interval evaluation can be used as a criterion to reject a box: if the evaluation of both sides of an (in)equation are inconsistent, the box has no solution. The filtering therefore reduces the facets of the box until each facet is evaluation-consistent.

4.2   Reducing Variable Redundancy

Interval propagation is *safe* in that the reduced box contains all the solutions to the initial one, but not *tight* in general: the obtained box over-approximates the set of solutions. This over-approximation is especially loose when variables have *multiple occurrences* in the constraints, due to a well-know phenomenon of variable decorrelation (interval propagation and interval evaluation behave as if every occurrence of the same variable denoted a different object). For instance, the natural interval evaluation of function $x - x$ with $I_x = [0, 100]$ gives $[-100, 100]$, which is rough estimation of its actual range, *i.e.*, $[0, 0]$.

The constraints which describe equilibria have a high number of variable redundancies. A natural way to improve the encoding is therefore to reduce this redundancy by expressing them in a syntactically improved way. We have investigated the two following improvements:

One-Sided Encoding. One source of redundancy in the previous formulation of maximality constraints was that very similar sums of monomials were repeated on both sides of the inequalities. This can easily be fixed by the following reformulation, which simply reported a -1 adjustment to every variable corresponding to the considered pure strategy:

$$\bigwedge_{i\in 1..n} \bigwedge_{j\in S_i} \left(\begin{array}{l} \displaystyle\sum_{s\in S_1\times...\times S_n} p_i(s).\pi^{ij}(s) \geq 0 \\[1em] \text{with :} \\[0.5em] \bullet\ \pi^{ij}(s) = \prod_{k\in 1..n} coef(k) \\[0.5em] \text{where } coef(k) := \begin{cases} \sigma_k[s_k] - 1 & \text{if } k = i \text{ and } s_k = j \\ \sigma_k[s_k] & \text{otherwise} \end{cases} \end{array} \right)$$

Once again, an example gives a better understanding of the trick (same example of eq. 2, we leave the 0 values explicit for better clarity); the constraint imposing that the payoff at the equilibrium be larger than on pure strategy a is:

$$\left(\begin{array}{l} 2\ \sigma_1[a]\ \sigma_2[\alpha] + 0\ \sigma_1[a]\ \sigma_2[\beta] \\ +\ 0\ \sigma_1[b]\ \sigma_2[\alpha] + 1\ \sigma_1[b]\ \sigma_2[\beta] \end{array} \right) - \left(\begin{array}{l} 2\ \underline{1}\ \sigma_2[\alpha] + 0\ \underline{1}\ \sigma_2[\beta] \\ +\quad 0 \qquad\quad +\ \ 0 \end{array} \right) \geq 0$$

This encoding can be rewritten into the following one, which avoids having to repeat twice similar sums of monomials:

$$\left(\begin{array}{l} 2\ (\sigma_1[a] - 1)\ \sigma_2[\alpha] + 0\ (\sigma_1[a] - 1)\ \sigma_2[\beta] \\ +\ 0 \qquad\quad \sigma_1[b]\ \sigma_2[\alpha] + 1 \qquad \sigma_1[b]\ \sigma_2[\beta] \end{array} \right) \geq 0$$

Factorised Encodings. — The second technique, which is a classical trick to reduce redundancy, is to factorise the polynomials. We use the observation that a polynomial of the form:

$$\sum_{\langle s_1,..,s_n\rangle\in S_1\times...\times S_n} p\langle s_1,..,s_n\rangle.\big(\sigma_1[s_1] * ... * \sigma_n[s_n] \big)$$

can be rewritten as

$$\sum_{s_1\in S_1} \sigma_1[s_1].\left(\sum_{\langle s_2,..,s_n\rangle\in S_2\times...\times S_n} p\langle s_1, s_2, ..., s_n\rangle.(\sigma_2[s_2] * ... * \sigma_n[s_n]) \right)$$

The transformation is recursively applicable, which directly gives a factorisation algorithm. As an example taken from one of our real (shortened) benchmarks, a polynomial whose syntax in the one-sided encoding is:

$$5.(x_{11}-1).x_{21}.x_{41} + 7.(x_{11}-1).x_{21}.x_{43} + $$
$$1.(x_{11}-1).x_{22}.x_{41} + 9.(x_{11}-1).x_{22}.x_{42} + $$
$$5.(x_{11}-1).x_{23}.x_{42} + 4.(x_{11}-1).x_{23}.x_{43} + x_{12}(...)...$$

can be encoded in a factorised form as follows:

$$(x_{11} - 1).\left(\begin{array}{l} x_{21} \cdot (5.x_{41} + 7.x_{43}) + \\ x_{22} \cdot (1.x_{41} + 9.x_{42}) + \\ x_{23} \cdot (5.x_{42} + 4.x_{43}) \end{array} \right) + x_{12}.\big(... \big) ...$$

Another factoring form has been experimented, where the factorisation is stopped at recursion depth one; in other words, when encoding the maximality constraints of a player, factorisation applies only to the variables belonging to this player.

5   Implementation

5.1   Description of the Implementation

Our approach was to use realpaver as a target language and to develop tools to translate games into its input format[3]. Three different encodings have been implemented: the one-sided one and its partially and completely factorised variants. The games are described using a simple syntax (*.nash* files in our experiment directories) which is exemplified hereafter.

```
(************* player 1 **************)
strategies 2 dependencies 1 2 :
                              (* input file for  the *)
          1  1  : 2,          (* 'bach-stravinsky'  *)
          1  2  : 0,          (*        game        *)
          2  1  : 0,
          2  2  : 1

(************* player 2 **************)
strategies 2 dependencies 1 2 :
    ...
    end
```

This set of tools readily enables a user to define and solve structured games by specifying the dependencies of each player and its payoff functions. To test the practical applicability of our approach, we have used a random generator of structured games, which is parametrised by the number of players, the maximal number of strategies for each of them, and coefficients for the *sparsity* (percentage of non-zero values of the payoff function) and the *connectivity* (percentage of other players which every player takes into account in its decisions) of the game[4].

5.2   Experimental Conclusions

Our experiments have dealt with more than 1000 instances of various sizes, each of which was translated using the 3 encodings. realpaver was run on all these games with different parameters, each time with a time limit of 60s. Our complete data sets, as well as the translation scripts, can be found on www.dis.uniroma1.it/~bordeaux/GAMES. Here is a summary of these results:

---

[3] In all our benchmarks, precision was set to 0.01, which corresponds to rough pseudo-solutions, but seems acceptable in comparison to other experiments found in the literature [20]. realpaver was used to find a unique solution; note that other choices are possible, and that it can also be used, for instance, to represent an enclosure of the solution set by a paving of arbitrary precision.

[4] We keep only the generated games whose dependency graph is connected and we allow a limited number of retries if it is not the case, hence the *generation failures* reported in some cases in our data.

Scalability: instances with around 10 variables can routinely be solved within the 60s limit. As soon as we go beyond this limit (*e.g.*, 5 players, each with 3 strategies each), the number of instances solved within the minute falls around the 50% (unsolved instances start to appear with 4 players and 3 strategies for some well-chosen sparsity/connectivity coefficients). Although some instances with 20 variables (and up to 5 players with 5 strategies) could be solved, they represent an almost negligible 5% of this size of instances.

This shows that non-trivial instances can be solved, which validates the applicability of a first interval-based approach, but a direct use of general-purpose interval methods does not compete with state-of-the-art methods based, for instance, on continuation [2]. Problem-specific interval propagation and search algorithms obviously have to be investigated in a near future. Indeed, if we consider the restrictions we have imposed to the tests (sparse, structured games, *low precision*), these results clearly indicate that games provide a source of challenging benchmarks for interval-based techniques, probably because of the fact that each constraint has multiple occurrences of almost all variables.

Choice of the Filtering Algorithm: a surprising pattern which clearly emerges is that basic interval propagation (hull consistency, HC in short) significantly outperforms box-consistency (or the mixed consistency technique used by default by realpaver), which contradicts conventional wisdom. A slowdown of 3 to 8 when using box instead of HC is typical; slowdowns of up to 100 are observed, and the number of instances where box behaves better than HC are scarce.

So far we cannot find an explanation to this phenomenon. HC is supposed to be especially inappropriate when the constraints have a high number of redundancies, which is the case here, but box does not seem to achieve a significantly better pruning, and is more time-consuming.

Impact of the Symbolic Representation: without any surprise, gains are obtained thanks to the factorised forms, with a typical speed-up of 2 (and up to 4 in rare cases) in comparison to the naive encoding. Nevertheless, this clearly does not allow to solve instances of a significantly larger size.

The best tuning we have found is the combination of hull consistency with a *largest first* strategy (`realpaver -hc4 -lf`), applied on the factorised encoding.

6   Conclusion and Perspectives

The amount of computer science and AI research related to game theory recently reported [7, 11, 8, 20, 4, 3, 2, 10, 9] shows that there is a high demand for tools for computing equilibria on n-person games. The starting point of our work was our conviction that the problem, which is tightly related to continuous constraint satisfaction, should receive more attention from the CP community. One goal of the paper was to give a CP-oriented and, we hope, accessible overview of the problem, and to provide readily available benchmarks and converters from games to CSP which can help other teams performing experimentations.

As a byproduct, we obtain generators of satisfiable yet hard polynomial instances for continuous solvers, which is of independent interest since the problem of generating continuous CSP has not been much addressed. Our experiments seem to reveal unusual results on the comparison of consistency techniques on randomly-generated data, and some work remains to be done to explain why they contradict conventional wisdom, with a box consistency typically less effective than basic interval propagation.

To the best of our knowledge, the application of interval constraint propagation for game-theoretic problems had not been considered prior to the investigation reported therein. On the one hand, the use of interval-based constraint solving techniques seems natural: since NASH equilibria are in general irrational numbers, approximating them with a certain precision might be the only viable approach in practice, and intervals seem the natural tool in this case. On the other hand, the high number of variable redundancies in the encodings of the problem makes it seemingly challenging for interval methods, and it is not clear so far whether interval methods can compete with continuation on this class of problems. Our approach was nevertheless naive since we directly used a general-purpose solver, and we hope that tailored propagation and search algorithms will lead to improvements.

Many techniques have been developed in the game-theory community to find equilibria [12], and yet tools which can consistently solve n-person games with satisfactory robustness and scalability are just starting to appear [11, 2, 17]. It is therefore clear that a long-term, collective effort is needed to obtain the gradual improvements which will eventually lead to efficient tools with *validated* numerical results, which only interval methods can obtain in general. Our hope is that this paper will motivate further research to explore, among other possible directions, cooperative algorithms. Since an essential subtask in computing equilibria is to determine the set of *supports* of the game, *i.e.*, the set of strategies which are given a non-zero probability, the problem also has a mixed combinatorial/real-valued flavour which makes it interesting from the viewpoint of both discrete and continuous CP, and mixed encodings might be another interesting direction.

Going one step further, it is clear that many exciting applications of CP to complex decision-making can emerge from the problems and tools proposed in the context of game theory, and that the particular normal-form games are far from the only case where CP and game theory can cross-fertilise. More general games (with incomplete information, in extensive-form), and other problems (mechanisms design, [19], *etc.*) will undoubtedly be investigated in a near future.

References

1. F. Benhamou and W. J. Older. Applying interval arithmetic to real, integer, and boolean constraints. *J. of Logic Programming (JLP)*, 32(1):1–24, 1997.
2. B. Blum, C. Shelton, and D. Koller. A continuation method for Nash equilibria in graphical games. In *Int. Joint. Conf. on Artificial Intelligence (IJCAI)*, pages 757–764. Morgan Kaufmann, 2003.
3. V. Conitzer and T. Sandholm. Complexity results on Nash equilibria. In *Int. Joint. Conf. on Artificial Intelligence (IJCAI)*, pages 765–771. Morgan Kaufmann, 2003.

Contents

4. R. S. Datta. Using computer algebra to find Nash equilibria. In *Int. Symp. on Symbolic and Algebraic Computation (ISSAC)*, pages 74–79. ACM, 2003.
5. Y. Gal and A. Pfeffer. A language for modeling agent's decision making processes in games. In *Int. Conf. on Autonomous Agents and Multiagent Systems (AAMAS)*, pages 265–272. ACM, 2003.
6. L. Granvilliers. An interval component for continuous constraints. *J. of Computational and Applied Mathematics*, 162(1):79–92, 2004.
7. M. J. Kearns, M. L. Littman, and S. P. Singh. Graphical models for game theory. In *Int. Conf. on Uncertainty in Artificial Intelligence (UAI)*, pages 253–260. Morgan Kaufmann, 2001.
8. D. Koller and B. Milch. Multi-agent influence diagrams for representing and solving games. In *Int. Joint. Conf. on Artificial Intelligence (IJCAI)*, pages 1027–1034. Morgan Kaufmann, 2001.
9. R. J. Lipton and E. Markakis. Nash equilibria via polynomial equations. In *South Amer. Symp. on Theor. Comp. Science (LATIN)*, pages 413–422. Springer, 2004.
10. R. J. Lipton, E. Markakis, and A. Mehta. Playing large games using simple strategies. In *ACM Conf. on Electronic Commerce (EC)*, pages 36–41. ACM, 2003.
11. M. L. Littman, M. J. Kearns, and S. P. Singh. An efficient, exact algorithm for solving tree-structured graphical games. In *Int. Conf. on Neural Information Processing Systems (NIPS)*, pages 817–823. The MIT Press, 2001.
12. R. McKelvey, A. McLennan, and T. Turocy. Gambit user manual, version 0.97.0.3. Technical report, The gambit project, 2003.
13. R. D. McKelvey and A. McLennan. Computation of equilibria in finite games. In *Handbook of Computational Economics*, chapter 2. North-Holland, 1994.
14. R. B. Myerson. *Game theory: analysis of conflict*. Harvard University Press, 1997.
15. J. F. Nash. Equilibrium points in n-person games. *Proc. of Nat. Academy of Science of the United States of America*, 36:48–49, 1950.
16. C. Papadimitriou. Algorithms, games, and the internet. In *ACM Symp. on Theory of Computing (STOC)*, pages 749–753. ACM, 2001.
17. R. Porter, E. Nudelman, and Y. Shoham. Simple search methods for finding a Nash equilibrium. In *US Conf. on Artificial Intelligence (AAAI)*, pages 664–669. AAAI Press, 2004.
18. S. Ratschan. Continuous first-order constraint satisfaction. In *Int. Conf. on AI and Symbolic Computation (AISC)*, pages 181–195. Springer, 2002.
19. T. Sandholm. Automated mechanism design: a new application area for search algorithms. In *Int. Conf. on Principles and Practice of Constraint Programming (CP)*, pages 19–36. Springer, 2003.
20. D. Vickrey and D. Koller. Multi-agent algorithms for solving graphical games. In *US Conf. on Artificial Intelligence (AAAI)*, pages 345–351. AAAI Press, 2002.
21. B. Von Stengel. Computing equilibria for two-person games. In *Handbook of Game Theory*, chapter 45. North Holland, 2002.

Constraint-Based Approaches to the Covering Test Problem*

Brahim Hnich, Steven Prestwich, and Evgeny Selensky

Cork Constraint Computation Center,
University College, Cork, Ireland
{brahim,s.prestwich,e.selensky}@4c.ucc.ie

Abstract. Covering arrays have been studied for their applications to drug screening and software and hardware testing. In this paper, we model the problem as a constraint program. Our proposed models exploit non-binary (global) constraints, redundant modelling, channelling constraints, and symmetry breaking constraints. Our initial experiments show that with our best integrated model, we are able to either prove optimality of existing bounds or find new optimal values for arrays of moderate size. Local search on a SAT-encoding of the model is able to find improved bounds on larger problems.

1 Introduction

Software and hardware testing play an important role in the process of product development. For instance, software testing may consume up to half of the overall software development cost [15]. Furthermore, even for simple software or hardware products, exhaustive testing is infeasible because the number of possible test cases is typically prohibitively large. For example, suppose we have a machine with 10 switches that have to be set, each with two positions. We wish to test the machine before shipping. Since there are 2^{10} possible combinations, it becomes impractical to test them all. Nevertheless, we might want only a small number of test settings such that every subset of, say three switches, gets exercised in all 2^3 possible ways. In such a case, the question becomes: How many test vectors do we need? This problem is an instance of the *t-covering array* problem.

A covering array $CA(t, k, g)$ of size b is an $k \times b$ array consisting of k vectors of length b with entries from $\{0, 1, \ldots, g-1\}$ (g is the size of the alphabet) such that the projection of any t coordinates contains all g^t possibilities. The objective consists in finding the minimum b for which a $CA(t, k, g)$ of size b exists. Covering arrays have been studied for their applications to drug screening and software and hardware testing. Over the past decade, there has been a body of work done in this field (See [1, 3–5, 11, 24, 25] for examples).

* The first author is supported by Science Foundation Ireland and an Ilog license grant. The third author is supported by Bausch&Lomb Ireland and Enterprise Ireland. This work has also received support from Science Foundation Ireland under Grant 00/PI.1/C075.

B. Faltings et al. (Eds.): CSCLP 2004, LNAI 3419, pp. 172–186, 2005.

Constructions for optimal covering arrays $CA(2, k, g)$ are known when the vectors are binary [17]. In [17], an exhaustive backtrack search is presented that is used to find new lower bounds on the sizes of optimal covering arrays $CA(2, k, g)$ where the alphabet Z_g is non-binary. However, in the general case the problem is NP-complete [11]. Most of the approaches use approximation methods where only upper and lower bounds of b are determined in polynomial time.

In this paper we propose modelling this problem, in its most general form, as a constraint program. We explore different models, and show that with a constraint programming approach we are able either to prove optimality of existing bounds, or to find new optimal values for problems of relatively moderate size. When the size of the problem increases our models' performance degrades, but we are able to find improved (though not necessarily optimal) bounds for larger problems by applying a local search algorithm to a SAT-encoding of the constraint model.

The rest of the paper is organised as follows. In section 2, we describe the covering test problem and give an overview of related work. Section 3 we detail the proposed constraint models. We then show how we extend our models to handle more general cases in Section 4. Section 5 presents our initial experimental results and a discussion. Finally, we conclude in Section 6 and outline our future directions.

2 The Covering Test Problem

The covering test problem[1] is a direct application of the problem of covering arrays arising in hardware and software testing[11].

Definition 1. Hartman and Raskin [11] *A covering array $CA(t, k, g)$ of size b and strength t, is a $k \times b$ array $A = (a_{ij})$ over $Z_g = \{0, 1, 2, ..., g-1\}$ with the property that for any t distinct rows $1 \le r_1 \le r_2 \le ... \le r_t \le k$, and any member $(x_1, x_2, ..., x_t)$ of Z_g^t there exists at least one column c such that $x_i = a_{r_i c}$ for all $1 \le i \le t$.*

Definition 2. Hartman and Raskin [11] *The covering array number $CAN(t, k, g)$ is the smallest b for which a $CA(t, k, g)$ of size b exists.*

The covering test problem is: for a given tuple $\langle t, k, g, b \rangle$ find a $CA(t, k, g)$ such that $CAN(t, k, g) = b$ or show that none exists. Informally, we wish to find a minimum number of test vectors, of k parameters each, over the alphabet Z_g such that the vectors contain all possible t-strings for every t-tuple of k parameters. Clearly, if $t = k$ and g is fixed then the number of test vectors is g^k and it is optimal. However, if $t < k$ then $\dot{g}^k$ is only an upper bound on the number of tests.

The problem of finding the minimum b can be solved iteratively by a series of constraint satisfaction problems with decreasing values of b. The solution with the smallest b is then guaranteed to be optimal.

---

[1] A description of this problem is also available as problem 45 in CSPLib, *www.csplib.org*

An approach to making software testing more efficient was presented by Cohen *et al.* [4], using test suites generated from combinatorial designs. The idea was firstly to identify parameters that induce the space of possible test scenarios; and secondly to select test scenarios so as to cover all the pairwise (or t-wise with $t > 2$ if necessary) interactions between the values of these parameters[2] This is analogous to earlier approaches [1, 24, 25]. A theoretical study [11] establishes properties of covering test suites, in particular the lower bounds on their size, and presents several ways to construct test suites to achieve the bounds asymptotically. The problem of minimising the number of test cases in a t-wise covering test suite for k domains of size n was, according to [11], first studied in [20].

Some papers consider the equivalent problem of maximising the number k of domains of size n in a t-wise covering test suite with a fixed number N of test cases [11]. This problem is referred to as *finding the size of the largest family of t-independent n-partitions of an N-set*. To determine the minimum number of test vectors for a t-wise covering of parameters with Boolean values is known to be NP-complete [23]. Related problems are finding a test suite with minimum deficiency given a fixed budget for executing a maximum of N tests, and a minimum test suite with a fixed relative deficiency (deficiency over the total number of t-subsets).

[16] discusses a practical issue of extending a given test suite to account for an additional parameter. The authors present an optimal algorithm for adding new rows to the test suite, once a new column has been inserted. However, their algorithms for adding a new column are either exponential or suboptimal. [3] presents a technique for reducing the covering test suite problem to graph-coloring. Even though this approach is more general, it is advantageous only for non-uniform coverage.

In [9, 12, 13] applications of covering suite generation are dealt with ranging from testing a satellite system to diagnosis in digital logic devices. This is known as the diagnosis problem and is generally solved via Built-In-Self-Testing (BIST). BIST is a relatively new area and is the leading approach in industrial testing. It offers low hardware overheads and quick testing capabilities [13]. The authors of [13] establish a link between BIST techniques and combinatorial group testing (CGT) [6]. They formulate the diagnosis problem, discuss the shortcomings of some contemporary BIST approaches, and overview standard CGT diagnosis algorithms such as *digging*, *multi-stage batching*, *doubling* and *jumping*. With these algorithms they achieve improvements over the BIST techniques, and present new hybrid diagnosis algorithms called *batched digging* and *batched binary search*.

To the best of our knowledge, no-one to date has looked at this area from a constraint perspective. Given the success of constraint technology in industrial combinatorial optimization, this paper is our first attempt to bridge this gap, and to see if constraint-based approaches can compete with existing methods.

---

[2] As [4] points out, the experience of Telcordia Technologies – formerly Bell Communications Research – is that pairwise coverage is sufficient for good code coverage and checking the interactions of system functions.

3 Constraint-Based Approaches

In this section we explore some models that exploit non-binary (global) constraints, redundant modelling, channelling constraints, and other features of Constraint Programming (CP). Many scheduling, assignment, routing and other decision problems can be efficiently and effectively solved by CP models consisting of matrices of decision variables (so-called "matrix models" [8]). We can model the problem of generating test vectors using multiple matrix models. Without loss of generality, in what follows we assume for clarity that we have a Boolean alphabet $Z_2 = \{0, 1\}$.

3.1 A Naive Matrix Model

As an example consider generating test vectors for all triples of 5 Boolean parameters ($t = 3$, $k = 5$, $g = 2$). The matrix in Figure 1 is a solution to this Boolean covering test problem, in which $b = 10$. Note that we highlight all possible combinations of 0 and 1 in the first three columns; this property holds for any triple of columns.

```
1 2 3 4 5
0 0 0 0 0
0 0 0 1 1
0 0 1 0 1
0 1 0 0 1
0 1 1 1 0
1 0 0 0 1
1 0 1 1 0
1 1 0 1 0
1 1 1 0 0
1 1 1 1 1
```

Fig. 1. A solution to the example.

A natural way to model the problem would be to introduce a $k \times b$ matrix of Boolean variables. However, we find it hard to express the *coverage* constraints, that is every t-parameters get combined in all possible 2^t ways. For every t-parameters in each row we introduce a Boolean variable for each combination that is set to true whenever these t-parameters cover that particular combination, by means of *reification constraints*. We then impose the constraint that each combination should occur at least once, using a sum constraint over the auxiliary Boolean variables.

Unfortunately, posing the coverage constraints on this matrix of decision variables introduces too many auxiliary variables and reification constraints. Furthermore, such a way of enforcing the coverage constraints makes constraint propagation inefficient and ineffective. We therefore need a different model where such coverage constraints can easily be expressed and propagated efficiently.

3.2   An Alternative Matrix Model

In our previous example, there are $\binom{5}{3} = 10$ triples of the original parameters:

$$T = \{\langle 1,2,3\rangle, \langle 1,2,4\rangle, \langle 1,2,5\rangle, \langle 1,3,4\rangle, \langle 1,3,5\rangle, \langle 1,4,5\rangle,$$
$$\langle 2,3,4\rangle, \langle 2,3,5\rangle, \langle 2,4,5\rangle,$$
$$\langle 3,4,5\rangle\}$$

We can exploit an alternative viewpoint of the problem to concisely express the covering constraints. We again introduce a matrix of decision variables. The b rows in this matrix represent a possible setting of the parameters. Each column however, represents one of the possible $t-$combinations (in T in our example). The domain of each variable is $\{0, ..., 2^t - 1\}$, or $\{0, ..., 7\}$ in this example.

In this new matrix, every entry is a problem variable that column-wise represents the above parameter triples T starting from left to right. The value 0 in this matrix stands for value combination $\langle 0,0,0\rangle$, 1 for $\langle 0,0,1\rangle$, and so on.

Coverage Constraints. Using the alternative matrix model, we can easily express the coverage constraints with the help of *global cardinality constraints* [19]. For each column we must guarantee that each value occurs *at least once* and *at most* $b - 2^t + 1$ times. This ensures that we cover all possible values of any t parameters.

Intersection Constraints. Because the variables in the first and the second column share digit positions 1 and 2 in the test vectors, the parameter values (0 or 1) in these positions should be the same. With the alternative model, we introduce the burden of expressing such *intersection* constraints. So for every row r and every two columns c_1 and c_2, if the two columns share some positions then we state a binary constraint between the variables (r, c_1) and (r, c_2) in the alternative matrix. For instance, for each row r the constraint between every two variables $M[r, \langle 1,2,3\rangle]$ and $M[r, \langle 1,2,4\rangle]$ in columns 1 and 2 can be expressed extensionally as follows:

$$\{\langle 0,0\rangle, \langle 0,1\rangle, \langle 1,0\rangle, \langle 1,1\rangle, \langle 2,2\rangle, \langle 2,3\rangle, \langle 3,2\rangle, \langle 3,3\rangle, \langle 4,4\rangle,$$
$$\langle 4,5\rangle, \langle 5,4\rangle, \langle 5,5\rangle, \langle 6,6\rangle, \langle 6,7\rangle, \langle 7,6\rangle, \langle 7,7\rangle\}$$

$\langle 1,2,3\rangle$	$\langle 1,2,4\rangle$	$\langle 1,2,5\rangle$	$\langle 1,3,4\rangle$	$\langle 1,3,5\rangle$	$\langle 1,4,5\rangle$	$\langle 2,3,4\rangle$	$\langle 2,3,5\rangle$	$\langle 2,4,5\rangle$	$\langle 3,4,5\rangle$
0	0	0	0	0	0	0	0	0	0
0	1	1	1	1	3	1	1	3	3
1	0	1	2	3	1	2	3	1	5
2	2	3	0	1	1	4	5	5	1
3	3	2	3	2	2	7	6	6	6
4	4	5	4	5	5	0	1	1	1
5	5	4	7	6	6	3	2	2	6
6	7	6	5	4	6	5	4	6	2
7	6	6	6	6	4	6	6	4	4
7	7	7	7	7	7	7	7	7	7

Fig. 2. The same solution as in Figure 1 but presented as an alternative matrix.

Note that the set of allowed tuples for such binary constraints differs depending on the type of the intersection. The tightness of such constraints [26] also varies. For example the tightness of the previous constraint is 0.25. Clearly, the tightness of the intersection constraints increases as the number of digit intersections decreases. There are situations when we have only one digit in common between a pair of variables (for instance, variables in the first column and variables in the 10th column). In that case, the constraint tightness is 0.5. Note also that we have b such constraints for every pair of tuples that intersect in at least one digit position.

3.3   An Integrated Model

In the naive matrix model we find it difficult to express the coverage constraints in such a way that we can reason efficiently and effectively about them. This is not the case with the alternative matrix model, where we can use global cardinality constraints for which efficient propagation algorithms exist [19]. The downside is that we have to explicitly express the intersection constraints.

In order to benefit from the effectiveness of each model, we propose integrating them by channelling the variables of the participating models. The disadvantages of this integration are the increased number of variables and additional channelling constraints to be processed. The advantage is, however, that we can easily state all problem constraints. We enforce the intersection constraints on the alternative matrix by simply channelling into the first matrix, benefiting at the same time from the efficient global cardinality constraints in the alternative model.

The channelling constraints associate each variable in the alternative matrix with t corresponding variables in the first matrix. The idea is to associate each possible way of combining the t parameters with a different value. For instance if $t = 3$ and the alphabet is binary then we constrain each variable ABC in the alternative matrix with its t corresponding parameters A, B, C as follows:

$$\langle ABC, A, B, C \rangle \in \{\langle 0, 0, 0, 0 \rangle, \langle 1, 0, 0, 1 \rangle, \langle 2, 0, 1, 0 \rangle, \langle 3, 0, 1, 1 \rangle$$
$$\langle 4, 1, 0, 0 \rangle, \langle 5, 1, 0, 1 \rangle, \langle 6, 1, 1, 0 \rangle, \langle 7, 1, 1, 1 \rangle\}$$

So for any t-covering we have $\binom{k}{t} \times b$ constraints of this type, and the arity of each constraint is $t + 1$.

3.4   Symmetry

A common pattern in matrix models is row and column symmetry [7]. A matrix has row symmetry and/or column symmetry when in any (partial) assignment to the variables, the rows and/or columns can be swapped without affecting whether or not the (partial) assignment satisfies the constraints. Clearly, any permutation of test vectors in a (non-)solution gives a symmetric (non-)solution. This means that the rows of our matrix models are indistinguishable and hence symmetric [7]. However, it is not trivial to see if the naive or the alternative matrix has column symmetry. The alternative matrix has no column symmetry while the naive (original) matrix does. Indeed, in the alternative matrix we associate each element with a particular combination of t parameters whereas in the naive matrix we do not distinguish where we project columns from as long as we make sure all parameter combinations are covered.

In Figure 3 the naive matrix (b) is the result of swapping columns 1 and 2 of the naive matrix (a). Both matrices represent symmetric solutions that correspond to $CAN(2,3,2)$. It is easy to see that because of the properties of the *covering constraints* that enforce every combination to occur at least once, such column swaps do not affect whether or not the (partial) assignment satisfies the constraints. Thus the naive matrix also has column symmetry.

The counterpart of such column symmetry in the alternative matrix is a complex combination of partial column symmetry and value symmetry among some variables. Note that the result of swapping the columns 1 and 2 in the naive model matrix (Figure 3) corresponds to the swap of columns 2 and 3 in the alternative matrix (c) and the application of the value symmetry that maps 0 to 0, 1 to 2, 2 to 1, and 3 to 3 to the variables of column 1 in (c), which results in matrix (d).

Thus, the naive matrix exhibits row and column symmetry, while the alternative matrix exhibits row symmetry and a complex form of symmetry (equivalent to the column symmetry in the naive matrix), but not column symmetry.

To break row and column symmetry in the naive matrix, we can order the rows and the columns lexicographically [7] using *lexicographic ordering constraints* [10]. Lexicographic ordering is a total order. Thus, by posing such an ordering constraint between every consecutive rows (columns), we break all row (column) symmetry [7]. Whilst it is easy to break all symmetry in one dimension of the matrix, breaking symmetry in both dimensions is harder, as the rows and columns intersect. After constraining the rows to be lexicographically ordered we distinguish the columns, thus the columns are no longer symmetric. Nevertheless, given a matrix with row and column symmetry, each symmetry class has *at least* one element where both the rows *and* columns are lexicographically ordered. Unfortunately, more than one element where both the rows and columns are lexicographically ordered may exist [7], so we cannot break all row and column symmetry. The lexicographic ordering constraint is linear in the size of the vector and it maintains *generalized arc consistency*.

(a)			(b)			(c)			(d)		
1	2	3	2	1	3	$\langle 1,2 \rangle$	$\langle 1,3 \rangle$	$\langle 2,3 \rangle$	$\langle 2,1 \rangle$	$\langle 2,3 \rangle$	$\langle 1,3 \rangle$
0	0	0	0	0	0	0	0	0	0	0	0
0	1	0	1	0	0	1	0	2	2	2	0
1	0	1	0	1	1	2	3	1	1	1	3
1	1	1	1	1	1	3	3	3	3	3	3
0	1	1	1	0	1	1	1	3	2	3	1
1	1	0	1	1	0	3	2	2	3	2	2

Fig. 3. Symmetric solutions corresponding to $CAN(2,3,2)$ represented as naive matrices (a,b) and as alternative matrices (c,d).

3.5   A Model for Local Search

Constraint solvers typically alternate variable assignment with constraint propagation; when propagation leads to an empty variable domain, backtracking occurs. An alternative way of finding solutions to constraint problems is local search. Usually starting from a randomly chosen assignment of all variables, single variables (or sometimes more than one) are selected and reassigned to a different value, each reassignment being a *local move*. The choice of variable and value is made heuristically, with no attempt to maintain completeness of search. This is in contrast to backtrack search, which is complete and can therefore find all solutions, or prove that no solutions exist.

The advantage of local search is that it can sometimes solve much larger problems than backtrack search. We decided to evaluate local search on our problem. We chose the Walksat algorithm, which has been successful on many problems and is publicly available. Walksat has several variants, and after some experimentation we selected the G variant [21], modified to break ties by preferring the variable that was flipped least recently (a well-known heuristic for improving search diversification). Walksat operates on Boolean satisfiability (SAT) models so we must first SAT-encode our problem. However, the best model for local search is not necessarily the best model for backtrack search [18]. Our SAT model is therefore not identical to the integrated matrix model.

As before we define a $k \times b$ matrix M of integers in Z_g. For each row i, column j and value s define a Boolean variable m_{ijs} which is true if s occurs in position (i, j) and false otherwise. We also define the alternative $\binom{k}{t} \times b$ matrix A of integers in Z_{g^t}. For each row i', column j' and value s' define a Boolean variable $a_{i'j's'}$. In the following constraints $1 \le i \le k$, $1 \le i' \le \binom{k}{t}$, $1 \le j, j' \le b$ and $1 \le s, s' \le g$. Each M and A position must take exactly one symbol:

$$\bigvee_s m_{ijs} \tag{1}$$

$$\bar{m}_{ijs} \vee \bar{m}_{ijs'} \tag{2}$$

$$\bigvee_s a_{i'j's'} \tag{3}$$

$$\bar{a}_{i'j's} \vee \bar{a}_{i'j's'} \tag{4}$$

where $s < s'$ in (2,4). The coverage constraints are:

$$\bigvee_{i'} a_{i'j's'} \tag{5}$$

To channel between the two matrices we infer the values of the t entries in M for the corresponding A entries:

$$\bar{a}_{i'js'} \vee m_{ijs} \tag{6}$$

for all i, i', j, s, s' such that $M_{ij} = s$ and $A_{i'j} = s'$ do not conflict. We refer to our SAT model as the *weakened matrix model* because it omits several constraints, as follows. Firstly the upper bound on the coverage constraints is hard to express in SAT. This is an implied constraint, and though implied clauses sometimes aid local search [2, 14] they are not a necessary part of the model. Secondly, symmetry breaking constraints can have a negative effect on local search performance [18]. Omitting them aids local search by increasing the number of SAT solutions, and also by reducing the size of the model and thus improving the flip rate (number of local moves per second). We therefore omitted upper bound and symmetry breaking constraints from our encoding.

The third difference is perhaps less obvious. When applying local search to a SAT-encoded CSP it is common to omit clauses ensuring that each CSP variable is assigned no more than one domain value [22], again improving performance. A CSP solution can still be extracted from a SAT solution by taking any one of the assigned domain values for each CSP variable. Here we may omit clauses (1,3,4). Note that we can still extract a CSP solution from any SAT solution: by clauses (5) in any SAT solution each combination of symbols occurs in at least one row of A for each combination of t columns; by clauses (6) each such occurrence induces the corresponding entries in M; and by clauses (2) no more than one value is possible in each M position. In fact the omitted clauses (1,3,4) are implied by clauses (2,5,6), and experiments suggest that omitting them makes little difference to the search effort. It reduces the size of the encoding but not its space complexity, which is dominated by the channelling constraints and is $O(\binom{k}{t} btg^t)$ literals.

3.6   Summary of the Models

We consider four matrix models for the covering test problem:

- **The Naive Matrix Model.** This model compactly represents the problem. However, it is difficult to express the coverage constraints in such a way that we can efficiently reason about them. This matrix has both row and column symmetry that we can efficiently and effectively reduce using lexicographic ordering constraints.
- **The Alternative Matrix Model.** This model overcomes the disadvantages of the previous model by the use of powerful global cardinality constraints. However, this comes at the cost of introducing the burden of expressing intersection constraints. This matrix has row symmetry that we can reduce using lexicographic ordering constraints. It has another complex form of symmetry that we do not know how to break efficiently and effectively.
- **The Integrated Matrix Model.** This model is an attempt to combine the complimentary strengths of both models. The coverage constraints are stated using the global cardinality constraints while the intersection constraints become redundant with the channelling constraints. We use the symmetry breaking constraints of the naive model as they are very efficient and effective. The overhead of this integrated model is the increased number of variables and the additional channelling constraints.
- **The Weakened Matrix Model.** This is a modification of the integrated matrix model, and designed for use with a SAT local search algorithm. It omits several constraints with the aim of increasing the number of SAT solutions and reducing runtime overheads.

4   Extensions

For reasons of clarity, we presented our models assuming a fixed binary alphabet and uniform coverage. However, our models can easily be extended to model different practical extensions:

- **Larger Alphabet.** To allow for larger alphabet, we need only to change the domain of the variables in both matrices. We also need to modify the channelling constraints. For instance suppose $t = 2$ and the alphabet is $Z_g = \{0, 1, 2\}$. The domain of the variable in the matrix in the naive model is Z_g while the domain of the variables in the alternative matrix is $\{0, \ldots, 3^2 - 1\}$, that is $\{0, \ldots, 8\}$. The channelling constraints between AB in the alternative matrix and its 2 corresponding parameters A and B in the first matrix become as follows:

$$\langle AB, A, B \rangle \in \{\langle 0, 0, 0 \rangle, \langle 1, 0, 1 \rangle, \langle 2, 0, 2 \rangle, \langle 3, 1, 0 \rangle, \langle 4, 1, 1 \rangle, \langle 5, 1, 2 \rangle, \langle 6, 2, 0 \rangle,$$
$$\langle 7, 2, 1 \rangle, \langle 8, 2, 2 \rangle\}$$

- **Heterogeneous Alphabets.** The model can easily be extended to allow heterogeneous alphabets. The domains of the variables as well as the channelling constraints need to be slightly changed to reflect this extension, but the essence of the models remains the same.
- **Partial Coverage.** To allow for partial coverage, we simply exclude those values that represent the combinations that need not appear in a solution from the global cardinality constraints.
- **Side Constraints.** Covering array problems can come with side constraints such as fixed columns or forbidden configurations [11]. CP is convenient for solving problems with such constraints, which can simply be added to the model.

5   Experiments

To evaluate the different models we ran a small set of experiments for a given alphabet, coverage strengths, and various parameter numbers k. First we report on backtracking experiments using a Pentium IV 1800 MHz 512 MB RAM machine running Ilog Solver 6.0.

In our experiments we used instances of the covering test problem with coverage strengths t of 3 and 4 over a Boolean alphabet $Z_2 = \{0, 1\}$. In each experiment we vary the size k of parameter vectors. Our initial experiments with the naive model showed that it was very inefficient and always outperformed by the other models. For this reason, we decided to exclude it from further analysis.

When using the alternative model we can only break row symmetry. However, with the integrated model we can break both row and column symmetry. Furthermore, we can break row symmetry either on the original or on the alternative matrix (not both), whereas column symmetry can be eliminated only on the original matrix. Our experiments demonstrated that the best strategy in terms of the amount of search and runtime, when using the integrated model, is to break row symmetry using the alternative matrix.

In the experiments we applied four different labeling strategies:

- *sdf-row*: Group the variables by rows from top to bottom, and for each row label the variable that has the smallest domain first. Assign the values in the lexicographic order;
- *sdf-col*: Group the variables by columns from left to right, and for each column label the variable that has the smallest domain first. Assign the values in the lexicographic order;
- *lex-row*: Group the variables by rows from top to bottom, and label each row lexicographically. Assign the values in the lexicographic order;
- *lex-col*: Group the variables by columns from left to right, and label each column lexicographically. Assign the values in the lexicographic order.

Experiments that we ran to determine $CAN(3, k, 2)$ for varying k showed that the best labeling heuristics were *lex-col* and *sdf-col* and that *lex-col* outperformed *sdf-col* and the other labeling heuristics on bigger instances. For example, using the alternative model with a time limit of 5 minutes, only *lex-col* could determine $CAN(3, 8, 2)$. Using the integrated model *lex-col* finds $CAN(3, 11, 2)$ in about 141 seconds, *sdf-col* in 281 seconds, whereas *lex-row* and *sdf-row* cannot find a solution.

Tables 1 and 2 display the results of the experiments in more detail. In the tables we use bold face to highlight the best result so far, whereas a star symbol means that the respective value is provably optimal. Our results also show that the integration of the different models is beneficial despite the increase in the number of variables. For instance, the best integrated model found the optimal value for $CAN(3, 8, 2)$ in around 22 seconds while the best alternative model in around 265 seconds. Note also that our results use the symmetry breaking constraints in all tested models. In fact, the symmetry breaking constraints play a vital part in the alternative and in the integrated models. For example, with the integrated model when we are solving the problem for $k = 5$ and $b = 10$ using *lex-col* labeling strategy together with row and column symmetry breaking we

Table 1. Alternative Model: Finding $b_{min} = CAN(3, k, 2)$ for different number of parameters k using the alternative model. Upper bounds UB on $CAN(3, k, 2)$ are taken from [11] for comparison. The runtime limit is 5 minutes, using *lex-col* (or *sdf-col*) as labeling heuristics.

k	b	Upper bound in [11]	runtime (sec)	soluble	no. of fails	no. of choice points
4	8*	8	0.01	+	28	29
5	8	12	0.03	−	32	31
5	9	12	0.19	−	161	160
5	10*	12	0.35	+	276	283
6	10	12	4.39	−	965	964
6	11	12	− (32.11)	−	− (6197)	− (6196)
6	12*	12	25.20	+	5003	5013
7	12*	13	128.34	+	9575	9591
8	12*	13	264.77	+	9575	9592

Table 2. Integrated Model: Finding $b_{min} = CAN(3, k, 2)$ for different number of parameters k using the integrated model. Upper bounds UB on $CAN(3, k, 2)$ are taken from [11] for comparison. The runtime limit is 5 minutes, using using *lex-col* as a labeling heuristic.

k	b	Upper bound in [11]	runtime (sec)	soluble	no. of fails	no. of choice points
4	8*	8	0.01	+	28	29
5	8	12	0.02	−	31	30
5	9	12	0.05	−	130	129
5	10*	12	0.10	+	186	191
6	10	12	0.67	−	625	624
6	11	12	4.67	−	3461	3460
6	12*	12	3.92	+	2642	2648
7	12*	13	13.11	+	4711	4721
8	12*	13	21.86	+	4714	4730
9	12*	18	75.47	+	10181	10205
10	12*	18	108.63	+	10185	10209
11	12*	18	140.90	+	10203	10230

Table 3. Values of $CAN(3, k, 2)$ and $CAN(4, k, 2)$ (**in bold** marked with a *) compared against the upper bounds from [11] (in parentheses); A time limit of 1 hour using the integrated model.

t	4	5	6	7	8	9	10	11
				k				
3	8* (8)	10* (12)	12* (12)	12* (13)	12* (13)	12* (18)	12* (18)	12* (18)
4	16* (16)	16* (24)	21* (28)	− (38)	− (42)	− (50)	− (50)	− (−)

obtain a solution in 0.10 CPU seconds with only 186 failures in contrast to over 5 CPU minutes and more than 640,000 failures when we break no symmetry. Finally, our approach proved optimality for $t = 3$ and $k \leq 11$ as well as helped improve the respective results in [11].

Encouraged by these initial results, we ran a further set of experiments to attempt solving $CAN(4, k, 2)$ for varying k. We observe in Table 3 that the best integrated model could find $CAN(4, k, 2)$ for $k \leq 6$ in 1 hour, and the improvements of the bounds that we obtained are significantly larger than the improvements we got on $CAN(3, k, 2)$. Overall, with the presented approach we can find provably optimal covering test suites for those instances which induce a moderate number of variables in our models. This translates to getting $CAN(3, k, 2)$ for up to $k = 11$ parameters (around 2000 variables) within a CPU time limit of 5 minutes. However, as problem size becomes larger the required amount of search proves computationally prohibitive.

With the aim of improving scalability we next applied the Walksat local search algorithm to the weakened matrix model, with infinite restart interval and noise parameter p set to values 0.2, 0.3 or 0.4. We ran Walksat on a 733 MHz Pentium III, using decreasing values of b until no solution was found after several minutes. The results for various values of t, k, g are shown in Table 4.

Table 4. Results for Walksat (W) and Hartman & Raskin [11].

t	3	3	3	3	3	3	3	3	3	3	4	4	4	4	3	3	3	3	3	4
k	9	10	11	12	13	14	15	16	17	18	7	8	9	10	5	6	7	8	9	5
g	2	2	2	2	2	2	2	2	2	2	2	2	2	2	3	3	3	3	3	3
b (W)	**12**	**12**	**12**	**15**	**16**	**17**	**18**	**18**	**18**	**20**	**24**	**24**	**24**	**24**	**33**	**33**	**40**	46	**51**	**81**
b [11]	18	18	18	18	19	19	19	19	24	24	38	42	50	50	45	45	45	**45**	75	135

Walksat was able to reproduce the improved bounds found by Solver, to further improve some bounds, and to solve several larger problems with better results than those of [11] (though they also give results for many larger instances). If proof of optimality is not required and we are not constrained to obtain a solution in a fixed time then local search is clearly a useful option. We do not expect the SAT approach to scale to much larger problems because of increasing SAT model sizes, but a local search algorithm using a higher-level constraint model could avoid this problem.

6   Conclusion

We presented constraint models of a core problem in combinatorial software testing: the covering test problem. We show that for moderate problem sizes with a CP approach one can find provably optimal solutions, which improves on the published results. We further showed that a local search algorithm on a SAT-encoding of the problem can find improved solutions for somewhat larger instances. These results show the applicability of constraint-based techniques to the problem, at least for instances up to a certain size. This approach may find application to less pure versions of the problem with side constraints, such as those found in some industrial applications. The easy handling of side constraints (simply by adding them to the model) is one of the advantages of CP.

In future work we will aim to further improve the presented results. One possible direction for improvement could be exploring the effects of different value ordering heuristics on backtrack search. Another direction is to design a dedicated local search algorithm for the problem; this would greatly reduce model sizes, which currently forms a bottleneck on the size of problems that we are able to solve.

References

1. S. Y. Boroday and I. S. Grunskii. Recursive Generation of Locally Complete Tests. *Cybernetics and Systems Analysis* 28:20–25, 1992.
2. B. Cha and K. Iwama. Adding New Clauses for Faster Local Search. *Proceedings of the Fourteenth National Conference on Artificial Intelligence*, American Association for Artificial Intelligence 1996, pp. 332–337.
3. C. Cheng, A. Dimitresku, and P. Schroeder. Generating Small Combinatorial Test Suites to Cover Input-Output Relationships. *Third International Conference On Quality Software (QSIC)*, USA, 2003, pp. 76–83.

4. D. M. Cohen, S. R. Dalal, M. L. Fredman, and G. C. Patton. The AETG System: An Approach to Testing Based on Combinatorial Design. *IEEE Transactions on Software Engineering* 23:437–444, 1997.
5. D. M. Cohen, S. R. Dalal, J. Parelius, and G. C. Patton. The Combinatorial Design Approach to Automatic Test Generation. *IEEE Software*, 1996, pp. 83–86.
6. D. Z. Du and F. K. Wang. *Combinatorial Group Testing and Its Applications.* World Scientific, 1991.
7. P. Flener, A. M. Frisch, B. Hnich, Z. Kızıltan, I. Miguel, J. Pearson, and T. Walsh. Breaking Row and Column Symmetries in Matrix Models. P. van Hentenryck, editor, *Proceedings of the Eighth International Conference on Principles and Practice of Constraint Programming*, 2002, pp. 462–476.
8. P. Flener, A. M. Frisch, B. Hnich, Z. Kızıltan, I. Miguel, and T. Walsh. Matrix Modelling: Exploiting Common Patterns in Constraint Programming. A. M. Frisch, editor, *Proceedings of the International Workshop on Reformulating Constraint Satisfaction Problems*, 2002, pp. 27–41.
9. G. Friedman, A. Hartman, K. Nagin, T. Shiran. Projected State Machine Coverage for Software Testing. *ACM SIGSOFT International Symposium on Software Testing and Analysis*, Roma, Italy, ACM Press, 2002, pp. 134–143.
10. A. M. Frisch, B. Hnich, Z. Kızıltan, I. Miguel, and T. Walsh. Global Constraints for Lexicographic Orderings. P. van Hentenryck, editor, *Proceedings of the Eighth International Conference on Principles and Practice of Constraint Programming*, 2002, pp. 93–108.
11. A. Hartman and L. Raskin. Problems and Algorithms for Covering Arrays. *Discrete Mathematics* 284:149–156, 2004.
12. J. Huller. Reducing Time to Market With Combinatorial Design Method Testing. *Proceedings of the 2000 International Council on Systems Engineering (INCOSE) Conference*, 2000.
13. A. B. Kahng and S. Reda. Combinatorial Group Testing Methods for the BIST Diagnosis Problem.*Proceedings of Asia and South Pacific Design Automation Conference*, 2004.
14. K. Kask and R. Dechter. GSAT and Local Consistency. *Proceedings of the Fourteenth International Joint Conference on Artificial Intelligence*, Morgan Kaufmann 1995, pp. 616–622.
15. N. Kobayashi. *Design and Evaluation of Automatic Test Generation Strategies for Functional Testing of Software*. PhD Thesis, Osaka University, 2002.
16. Y. Lei and K. C. Tai. In-Parameter Order: a Test Generation Strategy for Pairwise Testing. *Third IEEE High Assurance Systems Engineering Symposium*, 1998, pp. 254–161.
17. K. J. Nurmela. Lower Bounds on 2-Covering Arrays by Exhaustive Search. *Twenty-Fifth Australasian Conference on Combinatorial Mathematics and Combinatorial Computing*, 2000.
18. S. D. Prestwich. Negative Effects of Modeling Techniques on Search Performance. *Annals of Operations Research* 118:137–150, Kluwer Academic Publishers, 2003.
19. J.-C. Régin. Generalized Arc Consistency for Global Cardinality Constraints. *Proceedings of the Eighth National Conference on Artificial Intelligence*, 1996, pp. 25–32.
20. A. Renyi. *Foundations of Probability*. Wiley, New York, 1971.
21. B. Selman, H. Kautz, and B. Cohen. Noise Strategies for Improving Local Search. *Twelfth National Conference on Artificial Intelligence*, AAAI Press, 1994, pp. 337–343.

22. B. Selman, H. Levesque, and D. Mitchell. A New Method for Solving Hard Satisfiability Problems. *Tenth National Conference on Artificial Intelligence*, MIT Press, 1992, pp. 440–446.
23. G. Seroussi and N. H. Bshouty. Vector Sets for Exhaustive Testing of Logic Circuits. *IEEE Transactions Information Theory* 34:513–522, 1988.
24. D. T. Tang and C. L. Chen. Iterative Exhaustive Pattern Generation for Logic Testing. *IBM Journal of Research and Development* 28:212–219, 1984.
25. D. T. Tang and L. S. Woo. Exhaustive Test Pattern Generation With Constant Weight Vectors. *IEEE Transactions Computers* 32:1145–1150, 1983.
26. E. P. K. Tsang. *Foundations of Constraint Satisfaction*. Academic Press, 1993.

Super Solutions for Combinatorial Auctions*

Alan Holland and Barry O'Sullivan

Cork Constraint Computation Centre,
Department of Computer Science, University College Cork, Ireland
{a.holland,b.osullivan}@cs.ucc.ie

Abstract. Super solutions provide a framework for finding robust solutions to Constraint Satisfaction Problems [5, 3]. We present a novel application of super solutions to combinatorial auctions in which a bid may be disqualified or withdrawn after the winners are announced. We examine the effectiveness of super solutions in different auction scenarios that simulate economically motivated bidding patterns. We also analyze the drawbacks of this approach and motivate an extension to the framework that permits a more flexible and realistic approach for determining robust solutions.

1 Introduction

Many auctions involve the sale of various distinct items in which bidders perceive complementarity or substitutability between them. When auctioning such items, the auctioneer typically packages them in such a way as to maximize the complementarities amongst them and then puts these packages up for auction. When selling a farm, for example, it may be sold as a single item or divided into separate packages such as the farmhouse, outhouses, sites overlooking the beach, arable and non-arable land. When deciding on how to package the sale, it is impossible to know for sure what packages would maximize revenue. Bidders may view the complementarities between items differently and as the size of the auction grows it quickly becomes impossible for the auctioneer to know how the items should be packaged.

It is more economically efficient for the bidders to bid on combinations of items. Unfortunately the number of possible combinations of items of interest to each bidder grows exponentially as the number of items increases. The bids also need to be communicated to the auctioneer in a concise manner, which becomes increasingly difficult in large auctions.

The objective is typically to maximize revenue when selling items in a forward auction and to minimize cost when procuring items in a reverse auction. We consider the scenario where optimum revenue is non-essential but a *good robust* solution is essential. A *good* solution is one whose revenue is within a percentage of the optimum. A *good robust* solution is a good solution for which we know that even if winning bids are withdrawn another good solution can be found. Robust solutions are desirable in scenarios where suppliers or customers are unreliable and a good solution needs to be found with minimal upset to other bidders.

---

* This work has received funding from Science Foundation Ireland (Grant Number 00/Pl.1/C075) and from Enterprise Ireland, Research Innovation Fund (Grant Number RIF-2001-317).

B. Faltings et al. (Eds.): CSCLP 2004, LNAI 3419, pp. 187–200, 2005.

This paper is organized as follows. Section 2 introduces combinatorial auctions and describes the Winner Determination Problem. Section 3 presents super solutions and briefly discusses how they can be applied to combinatorial auctions. Section 4 outlines the experimental results achieved when finding super solutions with respect to various objectives one may wish to consider. It also tackles the optimization problems of finding a super solution with maximal robustness and revenue. Section 5 describes some of the limitations of super solutions and proposes an extension to the framework to overcome these difficulties. Some concluding remarks are made in Section 6.

2   Combinatorial Auctions

Combinatorial auctions fall into two categories, forward and reverse. In a reverse auction the auctioneer is seeking to procure goods and therefore minimize cost. If it is possible to purchase more items than are strictly necessary, the problem is known as a Set Covering Problem. The buyer may choose to stipulate in the auction rules that no surplus items are to be purchased. With the introduction of this constraint this problem becomes a Set Partition Problem.

In the remainder of this paper we focus our attention on forward auctions, where items are being sold and the objective is to maximize revenue. However, we will present our motivation for our interest in combinatorial auctions briefly below before considering forward auctions in more detail.

2.1   Motivation

The popularity of online auctions has increased in recent years because the internet promises to promote competition, thereby increasing revenue for the auctioneer. In an auction, where items exhibit complementarities or substitutabilities there exists a phenomenon known as the *exposure problem* [8]. This occurs when bidders seek a certain set of items but do not want to end up with a subset of the items that they may find valueless. This encourages cautionary bidding tactics that result in depressed bidding. Combinatorial auctions may alleviate the exposure problem by permitting bids on an arbitrary combination of items that suits the bidders needs. In this manner such auctions improve efficiency where items exhibit complimentarities/substitutabilties for the bidders.

Such auctions have been used in many real-world scenarios such as procurement for the Mars Corporation [6] and the sale of spectrum licences in America's Federal Communications Commission (FCC) auctions. The London Transport Authority also operated a combinatorial auction in their procurement of bus services from private operators [9, 8]. The Chilean government have also adopted combinatorial auctions for the supply of school meals to children. In the latter case, the quality of suppliers was considered as well as the bid amount in deciding the winner and the system also ensured there was no monopoly in any individual region. The reported supply costs have fallen by 22% since the adoption of the program [2].

In auctions where complementarities or substitutabilities are exhibited between items, there is a compelling argument for the introduction of combinatorial bidding to improve overall efficiency. Their application is spreading to other areas such as Supply Chain Management [12] that also demand robust solutions.

2.2   Forward Auctions: The Set Packing Problem

The auctioneer must determine the winner from all bids received. This is known as the *Winner Determination Problem* (WDP) and can be represented as a Set Packing Problem (SPP). In this case it is not necessary to sell all items in order to maximize revenue.

The SPP can be formulated as follows. Let I be the set of items to be auctioned and V is a collection of subsets of I. Let $x_j = 1$ if the jth set in V is a winning bid and c_j signifies the amount of that bid. Also, let $a_{ij} = 1$ if the jth set in V contains $i \in I$. The problem can then be stated as follows:

$$max \sum_{j \in V} c_j x_j$$

$$s.t. \sum_{j \in V} a_{ij} x_j \leq 1 \, \forall i \in I$$

We have encoded the SPP as a CSP where variables represent bids and the domains have only two values, 0 and 1, representing failure and success respectively. We have assumed the notion of *free disposal*[1]. Constraints between variables preclude the success of two bids containing the same item. The objective function is to maximize revenue and bid values are determined using a lookup table whose indices are variable-value pairs. An alternative formulation may include variables representing items whose domain consists of values representing the bids that include that item. Although when free disposal is assumed, not all items need to be assigned to bids so a null value is also included in the domain to indicate the item was unsold.

Example 1. Consider a simple example where an auctioneer is selling two items and there are four interested parties, bidders x_1, x_2, x_3 and x_4 (Table 1). Bidders x_1 and x_2 are interested in the first and second items respectively, bidding $0.60million for the relevant item. Bidders x_3 and x_4 seek both items only and bid $1.15million and $1.10million for the pair but $0 for each item individually[2]. The revenue maximizing solution for the auctioneer is to sell the items separately to x_1 and x_2 thus securing $1.2million. It is impossible for the auctioneer to know in advance whether combining the items in a single sale would be profitable or not. Combining both items in this case would have resulted in $0.05million of lost revenue, since the winning bid would have been for $1.15million rather than $1.2million. Instead combinatorial auctions allow the bidders decide on parcels of items that suit their needs thereby improving overall efficiency.

Example 1 may be encoded as a CSP by taking the four bids as variables with domains containing 0 and 1. Bids x_1 and x_3 cannot both succeed, since they both require item A, so a constraint is added precluding the assignment in which both variables take the value 1. Similarly, bid combinations x_2 and x_3, x_2 and x_4, and x_3 and x_4 cannot win simultaneously. Therefore, in this example the set of CSP variables, V,

---

[1] If the notion of *free disposal* is assumed, not all items need to be covered. In other words, there is no penalty for not selling some items.

[2] Such arbitrary complementarities amongst different bidders are often seen in property sales.

Table 1. Bids in Example 1.

Bidders	A	B	AB
x_1	**0.60**	0.00	0.00
x_2	0.00	**0.60**	0.00
x_3	0.00	0.00	1.15
x_4	0.00	0.00	1.10

would be as follows: x_1, x_2, x_3 and x_4 whose domains are all $\{0, 1\}$. The constraints are $x_1 + x_3 \leq 1$, $x_1 + x_4 \leq 1$, $x_2 + x_3 \leq 1$, $x_2 + x_4 \leq 1$ and $x_3 + x_4 \leq 1$. A lookup table, $a[i][j]$, is used to determine the amounts corresponding to variable values and the objective function, if we wish to optimize, is to maximize the sum of these amounts, $max \sum_{x_i \in V} a[i][x_i]$.

3   Super Solutions and Combinatorial Auctions

The purpose of finding super solutions is that if the solution is perturbed slightly, another good solution may be found by changing a limited number of other variables. An (a,b)-super solution is one in which at most a variables may lose their values and a repair solution may be found by changing at most b variables [5]. Only a particular set of variables in the solution may be subject to change and these are said to be members of the *break-set*. For each variable in the break-set, a *repair-set* is required that comprises the set of variables whose values may change to provide another solution.

In an auction, for example, bids may be retracted so are included in the break set. However, if we also used variables to represent the items, these may not break if items may not be withdrawn from an auction after the winners are announced. Robust solutions are particularly desirable for combinatorial auctions, as opposed to single-unit auctions, because bid withdrawals/disqualifications can leave the auctioneer facing the exposure problem that is faced by bidders in single unit auctions. The exposure problem refers to the situation where bidders are left holding a set of items that are essentially valueless. Ideally this set of items would complement items associated with some of the successful bids so the solution may be repaired easily or the items are valued highly by another bidder. In the case of a single-unit auction the solution is inherently robust because if a bid is withdrawn then the second highest bid for that item is chosen as a repair.

Example 2. Let us consider a simple example with two variables X and $Y \in \{0, 1\}$. There are three possible solutions $\langle 1, 1 \rangle$, $\langle 0, 1 \rangle$ and $\langle 1, 0 \rangle$. $\langle 1, 1 \rangle$ could be considered a $(1,0)$-super solution because if either of the variables breaks we are still left with a solution after making 0 changes. However, solutions $\langle 1, 0 \rangle$ and $\langle 0, 1 \rangle$ may be regarded as $(1,1)$-super solutions because we can always repair one variable if necessary to form another solution.

Similarly, some variables in the break set may not break depending on their value in the solution. Winning bids may be retracted, whilst retraction of losing bids is meaningless and has no effect on the solution. Super solutions need only be recorded for winning bids, hence assignments representing losing bids are said to have robust values.

Robust values aid the search for super solutions because we do not have to worry about finding repair solutions for the possible failure of those variables. In an auction scenario, solution failure is only likely to occur when one or more successful bids are withdrawn or disqualified. Consider Example 1, the optimal solution is $x_1 = 1$, $x_2 = 1$, $x_3 = 0$ and $x_4 = 0$. There are two winning bids involved in this solution so we only need to worry about repair solutions for variables x_1 and x_2. If x_3, for example, was withdrawn from the auction after the announcement of the winners, it would have no material effect on the solution. The WDP is an optimization problem that seeks to maximize revenue. Therefore, as well as being concerned with finding robust solutions, we also wish to find robust solutions with maximal revenue. The following example highlights this point.

Example 3. Consider the auction given in Example 1, with the additional constraint that we require a robust solution and we are willing to compromise on revenue but only by 20%. To be more precise, a (1,1)-super solution that matches the revenue constraint is required. This means that if any winning bid in the solution is withdrawn, a repair solution with revenue within 20% of optimum is necessary. In this case, if we were to choose the optimum solution $\langle 1, 1, 0, 0 \rangle$ we would be in trouble if either of the bids were withdrawn. Say x_1 was withdrawn or disqualified for some reason, the next best solution in terms of revenue is $\langle 0, 0, 1, 0 \rangle$. However, two variables would need to be changed in order to find a repair, therefore the optimum solution is not a (1,1)-super solution. However, $\langle 0, 0, 1, 0 \rangle$ is a (1,1)-super solution because a repair $\langle 0, 0, 0, 1 \rangle$ exists. This repair is in itself a super solution, but is dominated in terms of revenue by the previous solution therefore not chosen. Therefore, the best outcome in this situation is to select the (1,1)-super solution $\langle 0, 0, 1, 0 \rangle$, if we wish to find the maximal revenue for a robust solution within 20% of the optimum.

It is possible to reformulate the model so that the only solutions are super solutions [5, 3]. However, we use search because [5] provides evidence of its superior performance over reformulation. Finding super solutions can be computationally expensive [4] so a pure CP approach to the WDP that has an exponential search space does not scale very well. A hybrid approach incorporating OR techniques is required for larger auctions. Section 4 concentrates upon a fixed size of problem (20 items and 100 bids) and shows how (1,b)-super solutions are achievable for tighter revenue constraints in auctions with shorter bids[3]. We refer the reader to [4] for a discussion on the difficulty of computing super solutions compared to straightforward CSPs. Some hybrid techniques that could aid the scalability of solving combinatorial auctions are briefly outlined also.

4   Experimental Results

In this section we consider the effect of considering robustness in combinatorial auctions. We consider a number of aspects: finding (1,b)-super solutions, optimizing robustness and optimizing revenue. We used the Combinatorial Auction Test Suite (CATS) [7] to generate sample auction data. We generated sample auction problems in which there are 20 items for sale and 100 non-dominated bids[4] that produce 100 CSP variables. We

---

[3] Few items in each bid.

[4] The CATS flags included int_prices with the bid_alpha parameter set to 1000.

have examined smaller combinatorial auctions because a pure CP approach needs to be augmented with global constraints that incorporate OR techniques to increase pruning sufficiently so that thousands of bids may be examined. There are a number of ways to accelerate the search for super solutions in combinatorial auctions although this is not the focus of our work. Polynomial matching algorithms may be used in auctions whose bid length is very small, such as those for airport landing/take-off slots. Another technique may be to use an LP relaxation of the SPP to form an upper bound on potential revenue in sub-branches of the search tree. This is soluble in polynomial time and can therefore greatly improve performance. Such additional techniques, that are outlined in [10], can aid the scalability of a CP approach but our aim in these experiments is to focus upon the robustness of various auction distributions and consider the tradeoff between robustness and revenue. Our experiments use the EFC constraint solver [1] with additional super solution extensions developed by Hebrard et al [4, 5].

Twenty instances of each problem were used to generate average results. We used various distribution types that simulate different economically motivated auction scenarios. The `regions-npv` distribution is modelled on a scenario in which items are location dependent and complementarity is a function of the proximity of these items in 2-dimensional space, such as in a spectrum or real estate auction and valuations are distributed normally. In the `arbitrary-npv` distribution, complementarities may not be as universal as geographical adjacency with valuations being normally distributed again. Bidders view the complementarity of items slightly differently. The `scheduling` distribution simulates an auction for time slices on a resource in a distributed job-shop scheduling problem. The bids for this distribution type tend to be shorter, therefore there are a more combinations of possibly successful bids. This increases the difficulty of finding the optimal winner in the WDP but also increases the likelihood of being able to find a robust solution because of the increased availability of repair solutions.

4.1 Constraint Satisfaction

In this experiment we first solve the auction optimally. This can be done using an efficient IP solver such as ILOG's CPLEX package or CABOB[11]. We then stipulate a minimum percentage of optimum revenue that is acceptable and the maximum number of variables that can change, b, when forming a super solution. We then use the constraint-based solver to search for a satisfactory super solution. An IP approach to establishing such robust solutions would be extremely difficult to implement. The results are presented in Figures 1, 2 and 3.

The contours on the horizontal plane of the graphs indicate the gradient of the surface in the graph. This helps illustrate the rates of fall-off in the running time and success rate for the different distributions. It is evident from these contours that the `scheduling` distribution reaches a very high success rate with $b = 1$ and acceptable revenue of at least 90%, for example (Figure 3). The contours also help show where the peak running times are encountered for the various distributions.

These figures indicate the increasing levels of complexity for the various distributions: the `arbitrary-npv` being the easiest and `scheduling` distribution being the hardest. We can surmise from these graphs that when the constraints are tight, (b is low and minimal revenue is high), that it is easy to find that there is no super solutions

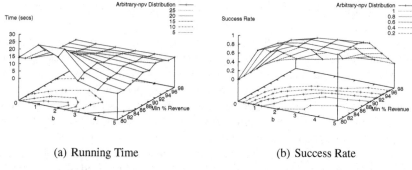

(a) Running Time   (b) Success Rate

Fig. 1. arbitrary-npv distribution – finding a (1,b)-super solution.

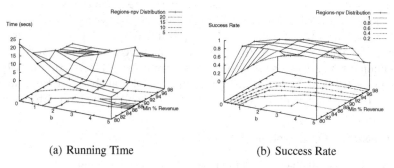

(a) Running Time   (b) Success Rate

Fig. 2. regions-npv distribution – finding a (1,b)-super solution.

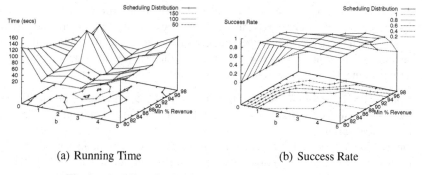

(a) Running Time   (b) Success Rate

Fig. 3. scheduling distribution – finding a (1,b)-super solution.

so running times are short. When constraints are very relaxed it is then easy to find a super solution. However, there is a phase transition area where the existence of a super solution is unpredictable and the running times are high. This peak in complexity is most clearly visible in Figure 3(a). The `arbitrary-npv` and `regions-npv` are evidently easier problems to solve. It is to be expected that running times increase with b because the search for a repair solution is longer. It can decrease in some instances when an increase in b results in many more solutions.

We can estimate from Figures 1, 2 and 3 that the hardest satisfaction problems for the various distributions occur when the success rate is approximately 75%.

It is more difficult to solve auctions in which there are many *short* bids (involving a small number of items) optimally because there are fewer constraints between the bid variables and deeper traversal of the search tree is required. We are seeking a robust solution within a given percentage of optimal revenue such that if any successful bid is withdrawn, a repair solution that is also within the same percentage of the optimum can be found by changing at most b other variables. In Figures 1, 2 and 3 we varied b from 0 to 5 and accepted the first super solution found. We did not consider values of b greater than 5 because most of the auction solutions contain 5 or less winning bids.

The `arbitrary-npv` distribution has long bids therefore few combinations of bids form valid solutions. This leads to fewer solutions and reduced time to solve. The `regions-npv` distribution has slightly shorter bids, therefore it is more difficult to solve than `arbitrary-npv` but easier than the `scheduling` distribution that has many short bids. The problem difficulty increases but the availability of robust solutions also increases because there are more possible repair solutions above the minimum threshold revenue. Figure 3(b) shows how the success rate for the `scheduling` distribution is better than for the `arbitrary-npv` and `regions-npv` distributions (see Figures 1(b) and 2(b) respectively). The increased availability of repair solutions accounts for the steeper contours towards 100% satisfiability in Figure 3(b).

4.2 Constraint Optimization

In a real-life scenario an auctioneer may seek a robust solution but would like the best robust solution in terms of either robustness or revenue. We may then employ an any-time algorithm that searches for the optimal super solution with maximum revenue being the objective function. We use a branch and bound algorithm that finds super solutions and optimizes on either reparability in the case of an over-constrained problem or revenue when there are many super solutions that satisfy the given constraints. This can be regarded as an anytime algorithm that finds the best possible robust solution in a given time-frame. Our analysis focuses on two forms of optimization:

1. Optimizing Robustness: Intended for use in an over-constrained scenario in which there are no super-solutions. We seek a solution that contains a minimal number of irreparable variables.
2. Optimizing Revenue: Intended for use in an under-constrained scenario in which there are many super solutions. We seek the super-solution with maximal revenue.

Optimizing Robustness. A scenario may occur where there exists no (1,b)-super solution that satisfies the minimal revenue criterion. We may seek a solution that minimizes the number of irreparable variables in a super solution thus compromising b but maintaining the revenue constraint. Hebrard et al. [4] have developed a (1,b)-`super Branch&Bound` algorithm that will find a super solution with a minimal number of irreparable variables if no super solution satisfies the given value of b.

Figure 4 shows how many variables in the super-solution will not provide a repair solution with at most b variable values being changed to form a new solution that supports the revenue constraint. These results show that easier distributions to solve are

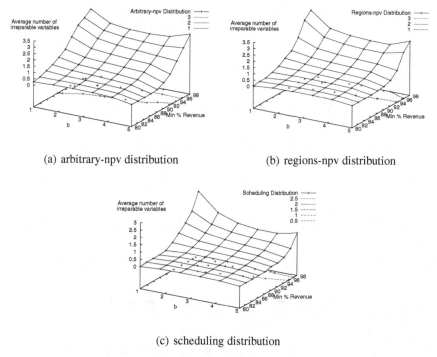

(a) arbitrary-npv distribution   (b) regions-npv distribution

(c) scheduling distribution

Fig. 4. Optimizing robustness.

less supportive of robust solutions[5]. When the revenue constraint on an auction with an `arbitrary-npv` distribution type is tight (95%) and few changes can be made to the solution ($b \leq 2$) then there are typically 2–3 bids in the auction that will not provide a repair solution given those constraints on revenue and b. The `regions-npv` and `scheduling` auctions have shorter bids and denser solution spaces so it is possible to find super-solutions that support such tight constraints.

Figure 5 compares the number of irreparable variables in the case of $b = 0$. Recall that a (1,0)-super solution is a solution in which if any winning bid is withdrawn, then the remaining winning bids still form a valid solution. This is only possible when there are many small winning bids whose value is less than that of the tolerable loss in potential revenue. This is a very tight constraint that is in fact unsatisfied by any of the sample auctions in our test-set. However, we can attempt to find a solution that minimizes the number of winning bids that do not satisfy this constraint, or *irreparable* variables. Figure 5 shows the increase in the minimum number of such variables as the constraint on minimum revenue is tightened. As this constraint is tightened, there is a trade-off against robustness.

There are two principle factors governing the reparability of a solution, the number of winning bids and the number of possible repair solutions. The `arbitrary-npv` distribution has fewer winning bids and fewer solutions so its reparability degrades

---

[5] Recall that auctions with long bids have fewer combinations of possible solutions so are easier to solve.

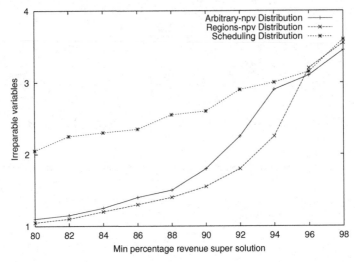

Fig. 5. Average number of irreparable variables in most robust solution (with $b=0$).

rapidly. The `regions-npv` and `scheduling` auctions tend to have more available repair solutions therefore degrade more slowly as the revenue increases. The latter distribution has more winning bids participating in the solutions. This accounts for why more of those bids do not have repair solutions. Also, there are more solutions in this distribution hence the increase in minimal revenue leads to a lower slope in the number of irreparable variables.

Optimizing Revenue. If there are many $(1,b)$-super solutions satisfying the revenue constraints then it is desirable to find the revenue-maximizing super solution. This a much more difficult problem than finding *any* super solution given a constraint on revenue.

We have developed a branch and bound algorithm that returns the optimal super-solution in terms of an objective function. In our case we search for a super solution whose revenue is maximal whilst the constraints on the revenue for repair solutions remain the same[6].

Figures 6(a), 6(b) and 6(c) show clearly that when we permit more variable changes, the expected increase in revenue of the optimal super solution increases significantly. Notice how optimization is far more difficult than satisfaction. This can clearly be seen by comparing the running times in Figure 7 with those in Figures 1, 2 and 3. We have restricted our analysis of the `scheduling` distribution to revenue greater than 90% because there are so many super solutions for revenue lower than this that optimization becomes extremely difficult. However, it is more difficult to see that as the constraint on acceptable revenue for repair solutions is tightened (minimum revenue for repair solutions increases) that in some cases this leads to a super solution of reduced optimal revenue. Note, however, that these graphs are averaged over those instances that proved satisfiable so as minimum revenue increases some problems became unsatisfiable therefore negating the decrease in revenue from satisfiable instances.

---

[6] An alternative approach may be to maximize the minimal revenue on the super solution and all repair solutions.

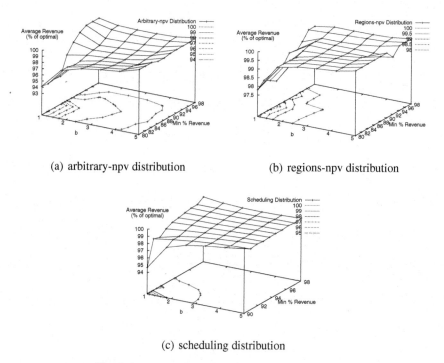

(a) arbitrary-npv distribution

(b) regions-npv distribution

(c) scheduling distribution

Fig. 6. Average optimal revenue of satisfiable instances.

5   Extensions to Super Solutions

Whilst the super solution framework provides an excellent framework for finding robust solutions, it is somewhat inflexible in some respects that are important to real-life applications. We have discovered some limitations in the approach when applied to combinatorial auctions. Firstly there is an underlying assumption that when a repair solution is created, the incurred cost of changing each variable's value in the repair set is the same and the total cost of repair is the cardinality of the repair set. In a real-world scenario, informing a losing bidder that they have now won because of the withdrawal/disqualification of a winning bid would typically incur less cost than informing a winning bidder that they have now lost. The auctioneer may have to break a contract or pay a penalty for such an action. This can be seen as a disadvantage of the super solution framework and militates against its deployment in real-life scenarios. Calculation of the cost associated with changing the losing/winning status of any bid is in reality a more complex issue that may depend on several other factors. Determining the legality of a repair solution by measuring the cardinality of the repair set may be overly restrictive in many application domains.

In other application domains such as scheduling, the cost associated with changing the value of a variable in a solution may depend on its destination value. Consider a factory scheduling problem where variables represent machines and values correspond to states. The cost of changing the state of any machine depends on both the source and destination states.

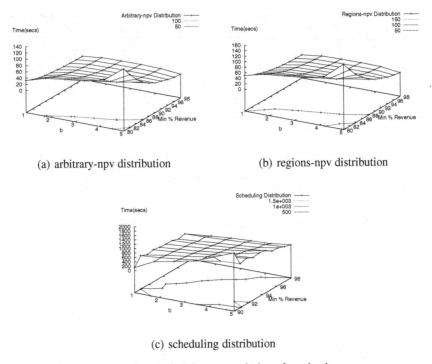

(a) arbitrary-npv distribution

(b) regions-npv distribution

(c) scheduling distribution

Fig. 7. Average time to find the super solution of maximal revenue.

The cost of changing a variable may also depend on the variable(s) that caused the break. For example, if a particular agent withdraws a bid from an auction, the auctioneer may favor rejection of the agents' other successful bids rather than disturbing an innocent party. The cost of a repair solution should therefore depend on the destination value and the breakage variable(s).

One possible approach is to extend the super solution framework to take account of the cost of repair. For example, we can use the concept of *inertia* as a metric for the cost of repair required to form an alternative solution. Previously the cardinality of the repair set was used to measure the cost of repair. We argue that changing some variables in a repair solution incurs less cost than others thereby motivating the introduction of a different metric for determining the legality of repair sets. The *inertia* of each value may be viewed as a measure of its aversion to change, therefore can be used to determine the cost of repair in the repair solution. We motivate this approach by using robust solutions to combinatorial auctions as an example application domain.

Hebrard et al [4] also described how some variables may fail (such as machines in a job-shop problem) and others may not. If we generalize this approach so that there is a probability of failure associated with each variable value, we can then alter the criteria for repair solutions according to each individual potential break. This probabilistic approach may be further enhanced by incorporating probabilistic failure distributions over time. Such distributions are common in reliability engineering and are often used to determine the mean time to failure. We may use them to maximize the mean time to *irreparable* failure, thus forming a robust solution for as long as possible when eventual failure is inevitable.

This extended feature of super solutions is motivated by the maintenance of robust solutions in combinatorial auctions. Agents' bids may be regarded as variables and Hebrard *et al's* (a,b)-super solution guarantees that an alternative solution may be found if a bids are retracted by changing at most b other variables. Maintaining a record of the reliability of assignments in the break set can help find repair solutions of lower *inertia* for values that are more likely to fail. All bids may not be viewed equally by the auctioneer and preferences may be shown for some agents. Eliminating a certain bid from the solution because of the retraction of another bid may incur varying costs for each bid, therefore the *inertia* associated with a value representing a winning bid reflects the unwillingness of the auctioneer to let that bid lose due to the actions of another. We also need to improve the model to include a constraint on the revenue in sub-branches of the search tree so that there is a tighter upper bound on revenue by using the LP relaxation of the problem. This would improve the scalability of the model.

Losing bidders typically would not mind being told that their bid was now accepted due to a winning bid's retraction so the *inertia* associated with such bids may be lower. This may be seen as a generalization of super solutions whose values of inertia are all 0 or 1. Furthermore, the *break-set* is regarded as the set of variables that may break, therefore the *inertia* of these variables values are all 1. Those variables not in the break-set have coefficients of inertia of 0 for all their values.

The overall cost of repair required to move from one solution to another is a function of the inertia on all values that need to be changed. An example of such a function could be the sum of the inertial values although not exclusively so. Taking this approach further, repairs may be made that restrict the impact on particular variables. In an auction scenario, we may wish that a repair solution does not impact too unfairly on any particular agent when another agent retracts a bid. The development of this extension to super solutions is our current topic of research.

6   Conclusion

Combinatorial auctions are becoming and increasingly popular means of selling/procuring items because they provide enhanced economic efficiency over traditional single unit auctions. Super solutions offer a promising platform for developing robust solutions for such auctions that may leave the auctioneer with an exposure problem if bids are withdrawn or disqualified. We have demonstrated the computational feasibility of finding super solutions for economically-motivated auction problems and shown how an anytime algorithm may find the best possible super solution given a limited time frame.

We also presented some limitations of the approach and suggested an extension to the framework. This extension incorporates a more flexible metric, termed *inertia*, for determining the legality of repair solutions and associating probabilities of failure to different variables. The introduction of weighted robustness of values complements this approach by allowing the construction of repair solutions of lower inertia for unreliable values. The development of this extension to super solutions is our current topic of research.

Acknowledgements

We are very grateful to Emmanuel Hebrard and Brahim Hnich for their assistance.

References

1. Fahiem Bacchus and George Katsirelos. EFC constraint solver.
   http://www.cs.toronto.edu/~gkatsi/efc/efc.html.
2. Rafael Epstein, Lysette Henríquez, Jaime Catalán, Gabriel Y. Weintraub, and Cristián Martínez. A combinational auction improves school meals in Chile. *Interfaces*, 32(6):1–14, 2002.
3. Matthew L. Ginsberg, Andrew J. Parkes, and Amitabha Roy. Supermodels and robustness. In *AAAI/IAAI*, pages 334–339, 1998.
4. Emmanuel Hebrand, Brahim Hnich, and Toby Walsh. Robust solutions for constraint satisfaction and optimization. In *Proceedings of ECAI 2004*, 2004.
5. Emmanuel Hebrard, Brahim Hnich, and Toby Walsh. Super solutions in constraint programming. In *Proceedings of CP-AI-OR 2004*, 2004.
6. Gail Hohner, John Rich, Ed Ng, Grant Reid, Andrew J. Davenport, Jayant R. Kalagnanam, Ho Soo Lee, and Chae An. Combinatorial and quantity-discount procurement auctions benefit mars, incorporated and its suppliers. *Interfaces*, 33(1):23–35, 2003.
7. Kevin Leyton-Brown, Mark Pearson, and Yoav Shoham. Towards a universal test suite for combinatorial auction algorithms. In *ACM Conference on Electronic Commerce*, pages 66–76, 2000.
8. Paul Milgrom. *Putting Auction Theory to Work*. Cambridge, March 2004.
9. Martin Pesendorfer and Estelle Cantillon. Combination bidding in multi-unit auctions. Harvard Business School Working Draft, 2003.
10. Tuomas Sandholm. Algorithm for optimal winner determination in combinatorial auctions. *Artificial Intelligence*, 135(1-2):1–54, 2002.
11. Tuomas Sandholm, Subhash Suri, Andrew Gilpin, and David Levine. CABOB: A fast optimal algorithm for combinatorial auctions. In *IJCAI*, pages 1102–1108, 2001.
12. W.E. Walsh, M.P. Wellman, and F. Ygge. Combinatorial auctions for supply chain formation. In *ACM Conference on Electronic Commerce*, pages 260–269, 2000.

Better Propagation for Non-preemptive Single-Resource Constraint Problems

Armin Wolf

Fraunhofer FIRST, Kekuléstr. 7, D-12489 Berlin, Germany
Armin.Wolf@first.fraunhofer.de

Abstract. Overload checking, forbidden regions, edge finding, and not-first/not-last detection are well-known propagation rules to prune the start times of activities which have to be processed without any interruption and overlapping on an exclusively available resource, i.e. machine. These rules are extendible by two other rules which take the number of activities into account which are at most executable after or before another activity. To our knowledge, these rules are based on approximations of the (minimal) earliest completion times and the (maximal) latest start times of sets of activities. In this paper, the precise definitions of these time values as well as an efficient procedure for their calculations are given. Based on the resulting time values the rules are re-formulated and applied to a well-known job shop scheduling benchmark.

1   Introduction

Job shop scheduling is one important representative of non-preemptive single-resource constraint problems which are in general NP-complete. However, recent publications present new efficient algorithms [3, 4, 12] for some known propagation rules. The application of these rules prune the potential start times of the activities to be processed and thus the search space for the solutions, i.e. the schedules.

In addition to other propagation rules like "edge finding" and others in [2, 10] a supplementary propagation rule is presented further restricting the earliest start times and latest end times of non-preemptive and time restricted activities requiring a single, exclusively available resource. This supplementary propagation rule follows the obvious observation that if there is an activity t and n other activities such that any k of them are not executable before (after) t then at most $k - 1$ are executable before (after) t and thus at least $n - k + 1$ activities have to be processed after (before) t. Any application of this rule requires the earliest completion times and latest start times of any sets of activities of known cardinality. In [2, 10] approximations of these values are used instead of the real values.

In this paper, we give a precise and efficiently computable definition of the earliest completion times and latest start times of sets of activities. Based on these definitions the minimal earliest completion times and the maximal latest start times of all subsets with cardinality k of any given set of activities is defined. In general, the use of resulting values yields better propagations than

B. Faltings et al. (Eds.): CSCLP 2004, LNAI 3419, pp. 201–215, 2005.

by the use of some approximated values. However, there are cases where the approximations for all subsets of activities perform better than the exact values only computed for the total set of all activities (cf. Example 2).

2   Non-preemptive Single-Resource Constraint Problems

Non-preemptive single-resource constraint problems are problems of finding a sequence for some non-interruptible time-restricted activities to be processed by the use of an exclusively available without overlapping in time. More formally, the problem is defined as follows:

Definition 1 (cf. [2]). *An activity t is non-interruptible and has a non-empty set of potential start times S_t, i.e. a finite integer set which is the domain of its variable start time $s(t)$. Furthermore, it has a fixed duration $d(t)$, i.e. a positive integer value*[1]*.*

Given a finite set of activities $T = \{t_1, \ldots, t_n\}$ with at least two elements ($n \geq 2$), the problem is to find a solution, i.e. some start times $s(t_1) \in S_{t_1}, \ldots, s(t_n) \in S_{t_n}$ such that either $s(t_i) + d(t_i) \leq s(t_j)$ or $s(t_j) + d(t_j) \leq s(t_i)$ holds for $1 \leq i < j \leq n$. Thus, a non-preemptive single-resource constraint problem is determined by a set of activities T which is either solvable *if there is such a solution or* unsolvable, *otherwise.*

In the following, we only consider non-preemptive single-resource constraint problems. Thus, it is always implicitly assumed that a set of activities T with at least two elements is given such that each activity $t \in T$ has a well-defined set of start times S_t and a well-defined duration $d(t)$. For each activity $t \in T$ a feasible start time $s(t) \in S_t$, i.e. a solution, has to be determined.

Furthermore, we identify for each activity $t \in T$ its *earliest start and completion times* $est(t)$ resp. $ect(t)$ as well as its *latest start and completion times* $lst(t)$ resp. $lct(t)$. Given the actual set of start times S_t of an activity $t \in T$ it holds $est(t) \leq \min(S_t)$, $lst(t) \geq \max(S_t)$, $ect(t) := est(t) + d(t)$, and $lct(t) := lst(t) + d(t)$. Especially before any application of a propagation rule let $est(t) := \min(S_t)$, $lst(t) := \max(S_t)$. Then, after any propagation the inequalities will hold.

It is easy to define the durations as well as the earliest start and completion times of sets of activities as canonical extension, i.e. given a non-empty subset of activities $M \subseteq T$ we define (cf. [2]):

$$d(M) := \sum_{t \in M} d(t) \qquad est(M) := \min_{t \in M}(est(t)) \qquad lct(M) := \max_{t \in M}(lct(t)) \ .$$

We might further define the earliest completion time (ect) and the latest start time (lst) of a non-empty subset of activities M as in [2, 10]:

$$ect(M) := est(M) + d(M) \quad \text{and} \quad lst(M) := lct(M) - d(M) \ .$$

However, these are only approximations of the real values as the following example shows:

---

[1] A generalisation with sets of potential durations that may be zero is possible, too.

Example 1. Considering the set of activities $\{a, b, u, v\}$ with $S_a := [0, 5]$, $S_b := [4, 5]$, $S_u := [7, 9]$ $S_v := [11, 15]$, $d(a) := 3$, $d(b) := 2$. $d(u) := 4$, $d(v) := 3$.

The earliest completion time of the activities $\{a, b\}$ is 6 which is greater than $est(\{a, b\}) + d(\{a, b\})$ which is 5.

The latest start time of the activities $\{u, v\}$ is 9 which is less than $lct(\{u, v\}) - d(\{u, v\})$ which is 11.

Thus any propagation based on the approximations might be improved by the use of the real values.

3   Earliest Completion Times of Sets of Activities

In the following, we define the earliest completion times of sets of activities and give some evidence that this declaration is sound.

Definition 2. *Given a non-empty set of activities* $T := \{t_1, \ldots, t_n\}$ *with* $n \geq 2$ *such that* $est(t_1) \leq \cdots \leq est(t_n)$ *holds. Now, let* $P := \{t_{i_1}, \ldots t_{i_k}\}$ *be a non-empty subset of* T. *We define* $P_{i_j} := \{t_{i_j}, \ldots, t_{i_k}\}$ *for* $j = 1, \ldots, k$. *This means that* $P = P_{i_1} \supseteq \cdots \supseteq P_{i_k} \neq \emptyset$ *holds. Furthermore, we define*

$$ect(P) := \max_{j=1,\ldots,k} est(P_{i_j}) + d(P_{i_j}) = \max_{j=1,\ldots,k} \left(est(t_{i_j}) + \sum_{l=j}^{k} d(t_{i_l}) \right)$$

to be the earliest completion time *of all activities in* P.

The declaration "earliest completion time" is justifiable: We assume that there is a solution such that all activities in P are completed before $ect(P)$. Then, it holds $s(t_{i_j}) + d(t_{i_j}) < ect(P)$ for each $j = 1, \ldots, k$. Thus, by definition of $ect(P)$, there are indices $m, p \in \{1, \ldots, k\}$ such that $ect(P) = est(t_{i_m}) + \sum_{l=m}^{k} d(t_{i_l})$ and $s(t_{i_p}) + \sum_{l=m}^{k} d(t_{i_l}) < ect(P)$ because t_{i_p} is scheduled first in the set P_{i_m}. It follows that $s(t_{i_p}) < est(t_{i_m})$ must hold and thus that $est(t_{i_p}) < est(t_{i_m})$ holds, which contradicts the given order, i.e. $est(t_{i_p}) \geq est(t_{i_m})$.

Obviously, ect is monotonic, i.e. if $N \subseteq M \subseteq T$ holds for two sets of activities N and M, then $ect(N) \leq ect(M)$ holds, too. Furthermore, $est(p) < ect(P)$ holds obviously for each activity $p \in P$.

Now, if we want to apply the propagation rule described in the introduction we are especially interested in the *minimal earliest completion time* of all sets of activities with cardinality k:

Definition 3. *Given a non-empty set of activities* $T := \{t_1, \ldots, t_n\}$ *with* $n \geq 2$ *such that* $est(t_1) \leq \cdots \leq est(t_n)$ *holds. We define:*

$$\mathsf{minSet}(T, 1) := \{t_i^{(1)}\}$$

where $t_i^{(1)} \in T$, *such that* $ect(\mathsf{minSet}(T, 1))$ *is minimal. If there is more than one such* $t_i^{(1)}$, *we choose the one with the smallest index* i.

For $k = 1, \ldots, n-1$ we define recursively[2]:

$$\mathsf{minSet}(T, k+1) := \mathsf{minSet}(T, k) \uplus \{t_i^{(k+1)}\}$$

where $t_i^{(k+1)} \in T \setminus \mathsf{minSet}(T, k)$, such that $ect(\mathsf{minSet}(T, k+1))$ is minimal. If there is more than one such $t_i^{(k+1)}$, we choose the one with the smallest index i.

Obviously, for $k = 1, \ldots, n$ the cardinality of the set $\mathsf{minSet}(T, k)$ is k. In the following we prove that $\mathsf{minSet}(T, k)$ has the *minimal* earliest completion time of all subsets of T with cardinality k. However, therefore it is necessary to prove for a subset $N \subseteq T$ of cardinality $k + 1$ ($k \in \{1, \ldots, |T| - 1\}$) containing the activity t that $ect(N) \geq ect(\mathsf{minSet}(T, k) \cup \{t\})$ holds. This requires the following lemma:

Lemma 1. *Let two sets of activities $N \subset T$ and $M \subset T$ be given such that there is an activity $t \in T \setminus (N \cup M)$. Further, let it holds*

$$ect(N) \geq ect(M) \ ,$$
$$est(t) \geq est(n) \quad \text{for each activity } n \in N,$$
$$est(t) \geq est(m) \quad \text{for each activity } m \in M.$$

Then, it holds $ect(N \uplus \{t\}) \geq ect(M \uplus \{t\})$.

Proof. By definition of ect it holds

$$ect(N \uplus \{t\}) = \max(ect(N) + d(t), est(t) + d(t))$$
$$ect(M \uplus \{t\}) = \max(ect(M) + d(t), est(t) + d(t)) \ .$$

Thus, the statement to be proved follows immediately.   □

Theorem 1. *Given a non-empty set of activities $T = \{t_1, \ldots, t_n\}$ with $n \geq 2$ such that $est(t_1) \leq \cdots \leq est(t_n)$ holds. Then, it holds*

$$ect(\mathsf{minSet}(T, k)) = \min_{N \subseteq T, |N| = k} ect(N) \qquad \text{for } k = 1, \ldots, |T|.$$

Proof. By Induction over k, we prove $ect(\mathsf{minSet}(T, k)) \leq \min_{N \subseteq T, |N| = k} ect(N)$. Then, the equality to be proven follows immediately, because $\mathsf{minSet}(T, k)$ is a subset of T with cardinality k.

Obviously, for $k = 1$ the statement holds by definition of $\mathsf{minSet}(T, 1)$.

Now, let N be an arbitrary subset of T with cardinality $k + 1$. Then, there is an activity $t_j \in N$ not contained in $\mathsf{minSet}(T, k)$. We choose t_j such that its index j is maximal, i.e. all other activities in N with greater index are also in $\mathsf{minSet}(T, k)$. In the following let $N' := N \setminus \{t_j\}$ and for $i = 1, \ldots, n$ let $N_i := N \cap \{t_i, \ldots, t_n\}$ and $N'_i := N' \cap \{t_i, \ldots, t_n\}$.

By definition of $ect(N)$ it holds $ect(N) \geq est(t_j) + d(t_j)$. If we further assume that $est(t_j) \geq ect(\mathsf{minSet}(T, k))$ holds, it follows immediately that $ect(N) \geq ect(\mathsf{minSet}(T, k) \uplus \{t_j\}) \geq ect(\mathsf{minSet}(T, k+1))$ holds. Consequently, we assume for the rest of the proof that $est(t_j) < ect(\mathsf{minSet}(T, k))$ holds.

---

[2] We use the operator '$\uplus$' for set union to emphasise the union of *disjoint* sets.

If it holds $ect(N) = est(N_i) + d(N_i)$ for an activity $t_i \in N'$ with $i < j$, then it follows immediately

$$ect(N) = est(N_i') + d(N_i') + d(t_j) \geq ect(\mathsf{minSet}(T, k)) + d(t_j)$$
$$\geq ect(\mathsf{minSet}(T, k) \uplus \{t_j\})$$
$$\geq ect(\mathsf{minSet}(T, k + 1))$$

by induction hypothesis and by definitions and thus the statement to be proved.

Thus, we assume in the following that $ect(N) = est(N_i) + d(N_i)$ holds for an activity $t_i \in N'$ with $i > j$. Further, we choose t_i such that its index i is maximal and the assumption holds, i.e. $ect(N) = ect(N_i)$. According to the choice of t_i it holds $ect(N \setminus \{t_i\}) < ect(N)$, otherwise there would be an $l > i$ such that $ect(N \setminus \{t_i\}) = est(N_l) + d(N_l) = ect(N)$ holds. According to our choices of t_i and t_j it holds $N_i \subseteq \mathsf{minSet}(T, k)$ and thus $ect(N_i) \leq ect(\mathsf{minSet}(T, k))$. Consequently, it follows $ect(N \setminus \{t_i\}) < ect(N) = ect(N_i) \leq ect(\mathsf{minSet}(T, k))$ contradicting the induction hypothesis, because $|N \setminus \{t_i\}| = k$. Consequently, this case cannot happen for any activity t_i with $i > j$.

Finally, if $ect(N) = est(t_j) + d(j) + d(N_i')$ holds for an activity $t_i \in N'$, then by definition of $ect(N)$ its index i is the smallest index of the activity in N' (or N) which is greater than j. Furthermore, it holds $est(t_j) \leq est(t_i) \leq est(t_j) + d(t_j)$. Otherwise, if we assume that $est(t_i) > est(t_j) + d(t_j)$ holds, then it follows $est(t_i) + d(N_i') = est(N_i) + d(N_i) > est(t_j) + d(j) + d(N_i') = ect(N)$ contradicting the definition of $ect(N)$.

For all activities $t_h \in N$ with $h < j$ it holds $est(t_h) + d(N_h \setminus N_j) \leq est(t_j)$. Otherwise, if we assume that there is an activity $t_l \in N$ with $l < j$ such that $est(t_l) + d(N_l \setminus N_j) > est(t_j)$ holds, then it follows $est(t_l) + d(N_l) > est(t_j) + d(N_j) = est(t_j) + d(t_j) + d(N_i') = ect(N)$ contradicting the definition of $ect(N)$.

Let $N_{j+1} := \{t_{u_1}, \ldots, t_{u_v}\}$ which is a subset of $\mathsf{minSet}(T, k)$ according to the choice of t_j. Now, we consider $N \setminus N_j = N \cap \{t_1, \ldots, t_{j-1}\}$. Obviously, it holds $|N \setminus N_j| \leq k$, thus by induction hypothesis and definition it holds $est(t_j) \geq ect(N \setminus N_j) \geq ect(\mathsf{minSet}(T, |N \setminus N_j|))$ and $\mathsf{minSet}(T, |N \setminus N_j|) \subseteq \mathsf{minSet}(T, k)$. Furthermore, by definitions and Lemma 1 it holds

$$ect(\mathsf{minSet}(T, |N \setminus N_j|) \uplus \{t_j\}) \leq ect((N \setminus N_j) \uplus \{t_j\})$$
$$ect(\mathsf{minSet}(T, |N \setminus N_j|) \uplus \{t_j, t_{u_1}\}) \leq ect((N \setminus N_j) \uplus \{t_j, t_{u_1}\})$$

$$\vdots$$

$$ect(\mathsf{minSet}(T, |N \setminus N_j|) \uplus \{t_j, t_{u_1}, \ldots, t_{u_v}\}) \leq ect((N \setminus N_j) \uplus \{t_j, t_{u_1}, \ldots, t_{u_v}\}) \ .$$

Consequently, $ect(N) \geq ect(\mathsf{minSet}(T, k + 1))$, because $N = (N \setminus N_j) \uplus \{t_j\} \uplus N_{j+1}$ and $\mathsf{minSet}(T, k) \uplus \{t_j\} = \mathsf{minSet}(T, |N \setminus N_j|) \uplus \{t_j\} \uplus N_{j+1}$ and further $ect(\mathsf{minSet}(T, k) \uplus \{t_j\}) \geq ect(\mathsf{minSet}(T, k + 1))$. $\qquad\square$

Theorem 1 shows that the declaration minSet is adequately chosen, because it defines a set of activities such that its earliest completion time is minimal with respect to all other sets of activities with the same cardinality.

4   Latest Start Times of Sets of Activities

The definitions of the latest start times of sets of activities as well as the definitions of the maximal latest start times of all sets of activities with cardinality k are symmetrical to the definitions in the previous section. Therefore, we have to consider the non-empty set of activities $T := \{t_1, \ldots, t_n\}$ with $n \geq 2$ in such a way that $lct(t_1) \geq \cdots \geq lct(t_n)$ holds. Then, for any non-empty subsets $Q := \{t_{i_1}, \ldots t_{i_k}\} \subseteq T$ and $Q_{i_j} := \{t_{i_j}, \ldots, t_{i_k}\}$ for $j = 1, \ldots, k$ let

$$lst(Q) := \min_{j=1,\ldots,k} lct(Q_{i_j}) - d(Q_{i_j}) = \min_{j=1,\ldots,k} \left(lct(t_{i_j}) - \sum_{l=j}^{k} d(t_{i_l}) \right)$$

be the *latest start time* of all activities in Q.

Symmetrically, the declaration "latest start time" is justifiable and all the properties holding for the earliest completion times of sets of activities hold analogously for the latest start times, too.

According to $\mathsf{minSet}(T, k)$ there are $\mathsf{maxSet}(T, k)$ having *maximal latest start time* of all sets of activities with cardinality k:

Corollary 1. *Given a non-empty set of activities T. Then, it holds*

$$lst(\mathsf{maxSet}(T, k)) = \max_{N \subseteq T, |N|=k} lst(N) \qquad for\ k = 1, \ldots, |T|.$$

Proof. The proof is analogous to the proof of Theorem 1.   □

With these definitions and properties of minimal earliest completion times and maximal latest start times of sets of activities of fixed size we are able to re-formulate the propagation rules presented in [2, 10].

5   Propagation Rules

Considering a non-preemptive single-resource constraint problem determined by a set of activities T. The following two rules check whether for an activity t any other k activities in $T \setminus \{t\}$ are not executable before (after) t. If so, at least $|T| - k$ activities in $T \setminus \{t\}$ have to be processed after (before) t:[3]

$$\forall t \in T\ \forall k \in \{1, \ldots, |T| - 1\}\ :\ ect(\mathsf{minSet}(T \setminus \{t\}, k)) > lst(t)$$
$$\Rightarrow S'_t := S_t \cap (-\infty, lst(\mathsf{maxSet}(T \setminus \{t\}, |T| - k)) - d(t)] \qquad (1)$$

$$\forall t \in T\ \forall k \in \{1, \ldots, |T| - 1\}\ :\ lst(\mathsf{maxSet}(T \setminus \{t\}, k)) < lct(t))$$
$$\Rightarrow S'_t := S_t \cap [ect(\mathsf{minSet}(T \setminus \{t\}, |T| - k)), +\infty) \qquad (2)$$

In applications the precondition of rule (1) might be extended by $lct(t) > lst(\mathsf{maxSet}(T \setminus \{t\}, |T| - k))$ and of rule (2) by $est(t) < ect(\mathsf{minSet}(T \setminus \{t\}, |T| - k))$ to avoid redundant rule activations.

---

[3] For any activity t the primed set of potential start times S'_t identifies an update of this set, i.e. the effect of any pruning operation resulting in a subset of S_t.

In contrast to the rules presented in [2, 10] these rules only consider the whole set of activities T and not all the subsets of T. We assumed that the restricted consideration of T results in the same propagation quality, i.e. each propagation performed by the consideration of all subsets of T is performed by the consideration of the set T itself. However this is not valid, as the following example shows:

Example 2. We consider the set of activities $T = \{t_1, t_2, t_3, t_4\}$ with $est(t_1) = \cdots = est(t_4) = 0$, $lct(t_1) = \cdots = lct(t_3) = 10$, $lct(t_4) = 8$, $d(t_1) = 1$, $d(t_2) = d(t_3) = d(t_4) = 3$. An application of the rule with $t := t_4$ and the subset $M := \{t_2, t_3\}$ of T prunes t's set of potential start times S_{t_4} which is initially $[0, 5]$:

$$minSet(M, 1) = \{t_2\} \text{ with } ect(minSet(M, 1)) = 3$$
$$minSet(M, 2) = \{t_2, t_3\} \text{ with } ect(minSet(M, 2)) = 6 \ .$$

Thus, $ect(minSet(M, 2)) > lst(t_4) = 5$ triggers Rule 1 where

$$maxSet(M, 1) = \{t_2\} \text{ with } lst(maxSet(M, 1)) = 7$$

which results in $S'_{t_4} := [0, 7 - d(t_4)] = [0, 4]$. Now, if we consider the set $I := T \setminus \{t\} = \{t_1, t_2, t_3\}$ it holds

$$minSet(I, 1) = \{t_1\} \text{ with } ect(minSet(I, 1)) = 1$$
$$minSet(I, 2) = \{t_1, t_2\} \text{ with } ect(minSet(I, 2)) = 4$$
$$minSet(I, 3) = \{t_1, t_2, t_3\} \text{ with } ect(minSet(I, 2)) = 6 \ .$$

Thus, only $ect(minSet(I, 3)) > lst(t_4) = 5$ triggers Rule 1 where

$$maxSet(I, 1) = \{t_1\} \text{ with } lst(maxSet(I, 1)) = 9$$

which results in no pruning.

However, these rules are further improvable: therefore we take any given (partial) order of the activities into account. Obviously, an activity s is *before* another activity t if the latest start time of s is less than the earliest completion time of t, because s cannot be started after the completion of t. Consequently, we define the relation

$$before(s, t) \quad \leftrightarrow \quad lst(s) < ect(t)$$

for any two different activities $s, t \in T$, $s \neq t$.[4]

Thus, for any activity $t \in T$ we are able to determine the set of activities to be processed before respective after t:

$$A_t := \{s \in T \setminus \{t\} \mid before(s, t)\} \text{ and } \Omega_t := \{s \in T \setminus \{t\} \mid before(t, s)\} \ .$$

---

[4] This **before** relation is called *detectable precedence* relation in [11].

Given these two sets, we know that all feasible start times of t must be in the interval $[ect(A_t), lst(\Omega_t) - d(t)]$, i.e.

$$\forall t \in T : S'_t := S_t \cap [ect(A_t), lst(\Omega_t) - d(t)] \tag{3}$$

and that the activities in A_t are eventually processed before t and the activities in Ω_t after t. Efficient propagation based on rule (3) is recently published in [11]. Furthermore, more propagation is possible: We know that at least $|A_t|$ activities and at most $|T| - |\Omega_t| - 1$ activities are executable before t and that at least $|\Omega_t|$ activities and at most $|T| - |A_t| - 1$ activities are executable after t. This allows a refinement of the sets of activities with minimal earliest completion time and maximal latest start time with respect to its cardinality. – For any subset $M \subseteq T$ we define *extended* minSets and maxSets:

$$\mathsf{minSetExt}(M, T, |M| + 1) := M \cup \{t_i^{(1)}\}$$

where $t_i^{(1)} \in T \backslash M$, such that $ect(\mathsf{minSetExt}(M, T, |M|+1))$ is minimal. If there is more than one such $t_i^{(1)}$, we choose the one with the smallest index i considering the ascending order with respect to the *est*'s.

$$\mathsf{minSetExt}(M, T, |M| + k + 1) := \mathsf{minSetExt}(M, T, |M| + k) \uplus \{t_i^{(k+1)}\}$$

where $t_i^{(k+1)} \in T \setminus \mathsf{minSetExt}(M, T, k)$, such that $ect(\mathsf{minSetExt}(M, T, k + 1))$ is minimal. If there is more than one such $t_i^{(k+1)}$, we choose the one with the smallest index i considering the ascending order with respect to the *est*'s.

$$\mathsf{maxSetExt}(M, T, |M| + 1) := M \cup \{t_i^{(1)}\}$$

where $t_i^{(1)} \in T \setminus M$, such that $lst(\mathsf{maxSetExt}(M, T, |M| + 1))$ is maximal. If there is more than one such $t_i^{(1)}$, we choose the one with the smallest index i considering the descending order with respect to the *lct*'s.

$$\mathsf{maxSetExt}(M, T, |M| + k + 1) := \mathsf{maxSetExt}(M, T, |M| + k) \uplus \{t_i^{(k+1)}\}$$

where $t_i^{(k+1)} \in T \setminus \mathsf{maxSetExt}(M, T, k)$, such that $ect(\mathsf{maxSetExt}(M, T, k + 1))$ is maximal. If there is more than one such $t_i^{(k+1)}$, we choose the one with the smallest index i considering the descending order with respect to the *lct*'s.

Analogous to Theorem 1 and Corollary 1 the following proposition holds:

Proposition 1. *Given a non-empty set of activities T. Then, for any subset $M \subseteq T$ it holds*

$$ect(\mathsf{minSetExt}(M, T, k)) = \min_{M \subseteq N \subseteq T, |N|=k} ect(N) \qquad and$$

$$lst(\mathsf{maxSetExt}(M, T, k)) = \max_{M \subseteq N \subseteq T, |N|=k} lst(N)$$

for $k = |M| + 1, \ldots, |T|$. $\qquad\qquad\qquad\square$

Based on these equalities, we are able to further refine the propagation rules w.r.t to the (partial) order of the activities:

$$\forall t \in T \ \forall k \in \{|A_t| + 1, \ldots, |T| - |\Omega_t| - 1\} \ :$$
$$ect(\mathsf{minSetExt}(A_t, T \setminus (\Omega_t \cup \{t\}), k)) > lst(t)$$
$$\Rightarrow S'_t := S_t \cap (-\infty, lst(\mathsf{maxSetExt}(\Omega_t, T \setminus (A_t \cup \{t\}), |T| - k)) - d(t)] \quad (4)$$

$$\forall t \in T \ \forall k \in \{|\Omega_t| + 1, \ldots, |T| - |A_t| - 1\} \ :$$
$$lst(\mathsf{maxSetExt}(\Omega_t, T \setminus (A_t \cup \{t\}), k)) < lct(t)$$
$$\Rightarrow S'_t := S_t \cap [ect(\mathsf{minSetExt}(A_t, T \setminus (\Omega_t \cup \{t\}), |T| - k)), +\infty) \quad (5)$$

Again, in applications the precondition of rule (4) might be extended by $lct(t) > lst(\mathsf{maxSetExt}(\Omega_t, T \setminus (A_t \cup \{t\}), |T| - k))$ and of rule (5) by $est(t) < ect(\mathsf{minSetExt}(A_t, T \setminus (\Omega_t \cup \{t\}), |T| - k))$ to avoid redundant rule activations.

The correctness of these rules is obvious. However, a formal proof is postponed.

For any efficient pruning based on all these propagation rules we are interested in fast algorithms calculating for any activity $t \in T$ and any necessary k the (extended) minSets and maxSets as well as their ects and lsts respectively.

6 Implementation and Complexity

For the implementation of the presented propagation rules we use several linked lists which are represented by some integer arrays. We only illustrate their usage for the computation of the minSets, the computation of the other sets is quite similar.

Given a set of activities $T = \{t_1, \ldots, t_n\}$ such that $est(t_1) \leq \cdots \leq est(t_n)$ and an activity $t_j \in T$, we initialise the following integer arrays running from 0 to $n + 1$ to compute the minSets of cardinality k which are subsets of $T \setminus \{t_j\}$:

index:	0	1	...	$j-1$	j	$j+1$	...	n	n+1
ect:	$-\infty$	$ect(t_1)$	...	$ect(t_{j-1})$	$ect(t_j)$	$ect(t_{j+1})$	...	$ect(t_n)$	$+\infty$
est:	$-\infty$	$est(t_1)$	...	$est(t_{j-1})$	$est(t_j)$	$est(t_{j+1})$	...	$est(t_n)$	$+\infty$
d:	0	$d(t_1)$	...	$d(t_{j-1})$	$d(t_j)$	$d(t_{j+1})$	...	$d(t_n)$	0
next candidate:	1	2	...	$j+1$	—	$j+2$	...	n+1	—
successor:	n+1	n+1	...	n+1	n+1	n+1	...	n+1	n+1
predecessor:	0	0	...	0	0	0	...	0	0
block occupation:	0	0	...	0	0	0	...	0	0
gap sum:	0	0	...	0	0	0	...	0	0

The values "$-\infty$" respective "$+\infty$" represent the smallest respective largest integers. Furthermore, the line "—" stands for any negative integer, representing the "nil" or "null" value in lists.

More illustrative, the activities in T are organised in linked lists as shown in Figure 1 (there exemplary for a minSet): In the upper list the candidates ("dashed" and "bricked" boxes) are organised while in the lower double-linked

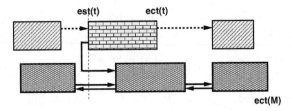

Fig. 1. The situation before the insertion of an activity.

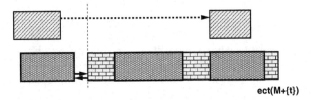

Fig. 2. The situation after the insertion of an activity.

list the activities of the optimal set of size $k-1$ are connected ("trellised" boxes). Initially, if $k = 1$ holds, this list is empty. In general, if $k > 1$, the earliest "blocks" occupied by these activities ("trellised" boxes) on the time line are connected. These blocks are the overlapping durations of the activities in the optimal sets. Each candidate is connected with a "successor" block for which the sums of all gaps between it and all its successors is known (initially, these sums are 0). With this information we are able to determine immediately how each candidate will change the *ect* when added to the optimal set of activities as shown in Figure 2 ("connecting bricks").

Thus, the computation of the minSets works as follows: Given the minSet with cardinality $k - 1$ we iterate over the current candidates for the extension of this set to an optimal set of size k, i.e. we compute how the addition of a candidate to the optimal set might increase its *ect*. For this purpose, we use the gap sums. After the iteration, we choose the candidate which increases the *ect* minimally. Then, this candidate is disconnected from the list of candidates and the list of occupied blocks is updated (cf. Figures 1 and 2). After the computations of all (extended) minSets and maxSets the rules presented in the previous section are applied straight forward.

For a better understanding of the core of the algorithm we show the computation of a minSet by an example:

Example 3. The following table shows a snapshot of a processing of the Fisher's & Thompson's job shop scheduling benchmark ft6 [5] during the consideration of the task with index 1 (its "next candidate" is valued "—"). The table shows that the linkage of the tasks is realised by using their indices. We use two "dummy" tasks (task #0 and task #7) representing the begin and the end of the linked lists. The list of candidates is linked via "next candidate". The list of "occupied blocks" is represented as a double-linked list and linked via "successor" and "predecessor". The line "—" represents the "nil" or "null" value in lists.

The following table contains the minSet with cardinality 2 consisting of the tasks #2 and #4 (the successor of #0 is #2, of #2 is #4, and of #4 is #7). The following figure shows that the *ect* of this minSet is 29, i.e. the minimal *ect* of all task sets with cardinality 2 not containing task #1. Further, the table and the figure shows that there is only one gap between task #2 and #4 with duration 2. Thus the gap sum for task #2 is 2 time units for all others it is zero.

index:	0	1	2	3	4	5	6	7
ect:	−∞	17	23	31	29	33	34	+∞
est:	−∞	12	13	23	25	27	27	+∞
d:	0	5	10	8	4	6	7	0
next candidate:	3	—	—	5	—	6	7	—
successor:	2	7	4	2	7	4	4	7
predecessor:	0	0	0	0	2	0	0	4
block occupation:	0	0	10	0	4	0	0	0
gap sum:	0	0	2	0	0	0	0	0

Further, we recognise that task #3 with successor #2 partially fits in this gap which will increase the *ect* to 35 as well as an extension of the minSet with task #5. Task #6 will increase the *ect* to 36, thus by convention the minSet is extended by task #3 resulting in a minSet of cardinality 3 with the minimal *ect* of all task sets with cardinality 3 not containing task #1.

The following table and figure shows the situation after an update of the data structures: the gap is filled and the durations of the tasks #2, #3 and #4 form a single block, also numbered #2, which occupies 22 time units.

index:	0	1	2	3	4	5	6	7
ect:	−∞	17	23	31	29	33	34	+∞
est:	−∞	12	13	23	25	27	27	+∞
d:	0	5	10	8	4	6	7	0
next candidate:	5	—	—	—	—	6	7	—
successor:	2	7	7	7	7	2	2	7
predecessor:	0	0	0	0	0	0	0	2
block occupation:	0	0	22	0	4	0	0	0
gap sum:	0	0	0	0	0	0	0	0

It should be noted that the data structures presented in the previous example are strongly motivated by the implementation of the propagation algorithm for the alldifferent constraint presented in [9].

In the remaining of this section, we show that the time complexity of the algorithm is cubic, i.e. $O(|T|^3)$:

Given a set of activities M with cardinality k the calculation of its ect or lst requires $O(k)$ computation steps, if we assume that an order of the activities with respect to their $ests$ and $lcts$ respectively is given. The sorting affords $O(k \log k)$ computation steps.

Given a (extended) minSet or maxSet with cardinality $k - 1$ we have to consider at most $|T| - k$ candidates for the extension of these sets to an optimal set of size k. Naively, this requires $O(k)$ calculations of $ects$ and $lsts$ and thus in total $O((|T| - k)k)$ computation steps. However, this is reduced to $O(|T| - k)$ computations steps due to the fact that the activities are organised in linked lists as shown in Figure 1 where each candidate is connected with a "successor" block for which the extension of all gaps between it and all its successors is known. With this information we are able to determine in $O(1)$ computation steps how each candidate will change the ect respective the lct when added to the optimal set of activities as shown in Figure 2.

Thus, for each $t \in T$ we have to compute $O(|T|)$ (extended) minSets and maxSets which require $O(|T|^2)$ computation steps: $O(|T|)$ computation steps for the appropriate candidate and $O(|T|)$ computation steps for an update of the linked lists of "occupied" blocks and its ect respective lst. Thus, the total time complexity is $O(|T|^3)$, i.e. the same as for the original rules [2, 10]. This also holds if the (partial) orders of the activities are taken into account: the calculation of the before relation affords naively $O(|T|^2)$ computation steps or $O(|T| \log |T|)$ computation steps in a more advanced setting (cf. [11]).

7   First Experiments

For first experiments with the propagation rules presented in the previous section, we implemented the necessary data structures and algorithms in our Java-based constraint solver firstcs [6]. In detail, we supplemented the implementations of the sweeping algorithms for forbidden regions (fr), edge finding with overload checking (efoc) and not-first/not-last detection (nfnl) presented in [12] with an implementation of the propagation rules (3), (4), and (5) (cf. Section 6). For a fix-point computation of the propagators we used the nested iteration shown in Figure 3: The optional propagation of these rules with cubic time complexity is performed after the more efficient sweeping algorithms having quadratic or even better complexity, i.e. $O(n \log n)$ [11]).

We applied these pruning algorithms to some well-known job shop scheduling benchmark optimisation problems: abz5, abz6 [7], ft10 [5], la16, la18, la20, la36 [8], and orb01-orb05 [1]. For all these problems it is sufficient to order the tasks on the machines totally to obtain a schedule for the given minimal makespan because forward scheduling will determine the activities' start times.

```
boolean hasChanged = true;
while (hasChanged) {
        while (hasChanged) {
                hasChanged = false;
                performEdgeFindingWhileSweeping();
                performNotFirstNotLastDetectionWhileSweepint();
                sweepOverForbiddenRegions();
        }
        if (BETTER_PROPAGATION) {
                performBetterPropagationWithBefore();
        }
}
```

Fig. 3. Combination and iteration over the pruning algorithms implemented in the constraint solver `firstcs`.

For that purpose a branching procedure with constraint propagation at each node of a search tree is used. The basic structure of the branching schema during this search process is as follows:

- select a machine with the highest relative demand. The relative demand is the ratio of the sum of all durations of the tasks on this machine and the difference between the latest completion and earliest start times of the task on this machine.
- select a task s on this machine with a smallest relative slack which is not ordered with respect to all other tasks. The relative slack is the ratio of the difference between the task's latest and earliest start time and its duration.
- select another task t on this machine a the smallest relative slack which is not ordered with respect to the task s.
- choose s before t. If this results in an inconsistency, choose t before s.
- repeat until all tasks on all machines are ordered totally or the search process fails finally.

We applied this search schema statically: the relative demands and slacks of the machines and their tasks are computed once at the beginning of the search process. Within the search, different kinds of constraint propagation are applied: forbidden regions, edge finding with overload checking, and not-first/not-last detection (fr+efoc+nfnl). Additionally, the propagation rules (3), (4), and (5) are applied, too (fr+efoc+nfnl+better-order). In the following table the different search steps are listed for different given makespans:

Table 1 shows that the additional pruning rules seem to result in better propagation of the single-resource constraints: the number of backtracks reduces significantly; in some case more than one order magnitude. However, the search is in none of the test cases faster. One reason is the prototypical implementation of the pruning rules presented in Section 6. Another, and more crucial reason is certainly the cubic time complexity of this implementation. Thus, we hope that further research will breed more efficient algorithms implementing the presented propagation rules.

Table 1. Experimental results showing the influence of the additional pruning rules.

problem	size	makespan	fr+efoc+nfnl		fr+efoc+nfnl+better-order	
			backtracks	runtime [msec.]	backtracks	runtime [msec.]
abz5	10 × 10	1,234	52,534	31,100	2,313	40,350
abz6	10 × 10	943	23,916	14,800	1,861	31,700
ft10	10 × 10	930	10,550	8,600	1,518	31,350
la16	10 × 10	945	702	937	53	1,850
la18	10 × 10	848	7,400	6,050	1,368	23,900
la20	10 × 10	902	5,863	3,850	1,421	17,750
la36	15 × 15	1,268	6,642	17,100	3,084	216,350
orb01	10 × 10	1,059	50,641	51,400	10,393	198,900
orb05	10 × 10	887	2,320	2,250	610	11,750

8   Conclusion and Future Work

In this paper the earliest completion and latest start times of sets of activities are precisely defined. Further, the sets of activities of fixed size with minimal earliest completion and latest start times with respect to their sizes are defined, too. Based on these definitions some well-known propagation rules are re-formulated: approximations of these values are replaced by their exact values. These rules are re-formulated twice: in the second step the tasks before and after a given task are also taken into account. Then, we explained how the necessary computations for an application of these rules are performed efficiently and showed by an example the used data structures. Finally, we applied these rules to a well-known benchmark problem affirming that the re-formulated rules reduces the number of search steps that are required to find a solution.

Future theoretical work will focus on more efficient algorithms reducing the current cubic time complexity. Practical work will focus on the improvement of the prototypical implementation and the execution of more experiments.

References

1. David Applegate and William Cook. A computational study of the job-shop scheduling problem. *ORSA Journal on Computing*, 27(3):149–156, 1991.
2. Philippe Baptiste, Claude le Pape, and Wim Nuijten. *Constraint-Based Scheduling.* Number 39 in International Series in Operations Research & Management Science. Kluwer Academic Publishers, 2001.
3. Nicolas Beldiceanu and Mats Carlsson. Sweep as a generic pruning technique applied to the non-overlapping rectangles constraint. In Toby Walsh, editor, *Proceedings of the 7th International Conference on Principles and Practice of Constraint Programming – CP2001*, number 2239 in Lecture Notes in Computer Science, pages 377–391. Springer Verlag, 2001.
4. Nicolas Beldiceanu and Mats Carlsson. A new multi-resource cumulatives constraint with negative heights. In Pascal van Hentenryck, editor, *Proceedings of the 8th International Conference on Principles and Practice of Constraint Programming – CP2002*, number 2470 in Lecture Notes in Computer Science, pages 63–79. Springer Verlag, 2002.

5. G. L. Thompson H. Fisher. Probabilistic learning combinations of local job-shop scheduling rules. In G. L. Thompson J. F. Muth, editor, *Industrial Scheduling*, pages 225–251. Prentice Hall, Englewood Cliffs, New Jersey, 1963.
6. Matthias Hoche, Henry Müller, Hans Schlenker, and Armin Wolf. firstcs – A Pure Java Constraint Programming Engine. In Michael Hanus, Petra Hofstedt, and Armin Wolf, editors, *2nd International Workshop on Multiparadigm Constraint Programming Languages – MultiCPL'03*, 29th September 2003. Online available at `uebb.cs.tu-berlin.de/MultiCPL03/Proceedings.MultiCPL03.RCoRP03.pdf`.
7. E. Balas J. Adams and D. Zawack. The shifting bottleneck procedure for job shop scheduling. *Management Science*, 34:391–401, 1988.
8. S. Lawrence. Resource constrained project scheduling: an experimental investigation of heuristic scheduling techniques (supplement). Technical report, Graduate School of Industrial Administration, Carnegie-Mellon University, Pittsburgh, Pennsylvania, 1984.
9. Alejandro Lopez-Ortiz, Claude-Guy Quimper, John Tromp, and Peter van Beek. A fast and simple algorithm for bounds consistency of the alldifferent constraint. In *Proceedings of the 18th International Joint Conference on Artificial Intelligence*, pages 245–250, Acapulco, Mexico, August 2003.
10. Wim P. M. Nuijten and Claude Le Pape. Constraint-based job shop scheduling with ILOG Scheduler. *Journal of Heuristics*, 3, 1998.
11. Petr Vilím. $o(n \log n)$ filtering algorithms for unary resource constraint. In *Proceedings of the International Conference on Integration of AI and OR Techniques in Constraint Programming for Combinatorical Optimisation Problems – CP-AI-OR'04*, number 3011 in Lecture Notes in Computer Science, pages 335–347, Nice, France, April 20–22, 2004. Springer Verlag, Heidelberg.
12. Armin Wolf. Pruning while sweeping over task intervals. In Francesca Rossi, editor, *Proceedings of the 9th International Conference on Principles and Practice of Constraint Programming – CP 2003*, number 2833 in Lecture Notes in Computer Science, pages 739–753, Kinsale, County Cork, Ireland, 30th September – 3rd October 2003. Springer Verlag.

Author Index